PENGUIN 1
ONE S

Neha Mehta is an Indian-Singaporean Fintech thought leader, lawyer, and award-winning entrepreneur. She is an Adjunct Professor with Nanyang Technological University, SP Jain and other global universities.

As the Founder and CEO of FemTech Partners, Neha amplifies the voice of Women in Tech and mentor's young women in Science, Technology, Engineering and Mathematics (STEM). FemTech Partners is a proud recipient of the British Chamber of Commerce Award. In the last seventeen years, Neha has worked with financial regulators in Asia, Oceania, and Europe to revamp policy, and infrastructure to include poorest of poor. Her work has contributed towards sustainable development goals of gender equality and financial inclusion through sustainable finance, and innovative digital financial services. Neha has worked with the UN, Deutsche Boerse, and various third-sector organizations.

Neha holds a double degree in Law and Commerce and attended King's College, London on a Chevening Fellowship. Neha is an alumna of the Fortune U.S. State Department Global Women Mentorship and an Austrian Leadership Fellow.

Neha is passionate about diversity, inclusion, and equity (DEI), and her pro-bono work has won her several awards including Globant Digital Disruptors Award, Kindness Leadership Award Asia Pacific, Global Future Fintech Leader, Stri Shakti Award, and Nav Shakti Award. A finalist for the STEM Trailblazer Baton Award, and International Inspiration Award.

Neha has been featured in coveted honours including Singapore 100 Women in Tech, FinTech top 50 and Asia's 50 Most Influential Women in Renewable Energy.

Neha is also a board member and advisor to impact led organizations. You can find out more at: Linktr.ee/nehamehta.

One Stop

Neha Mehta

PENGUIN BOOKS

An imprint of Penguin Random House

PENGUIN BOOKS

USA | Canada | UK | Ireland | Australia
New Zealand | India | South Africa | China | Southeast Asia

Penguin Books is part of the Penguin Random House group of companies
whose addresses can be found at global.penguinrandomhouse.com

Published by Penguin Random House SEA Pte Ltd
9, Changi South Street 3, Level 08-01,
Singapore 486361

First published in Penguin Books by Penguin Random House SEA 2023
Copyright © Neha Mehta 2023

ISBN 9789815127881

Typeset in Garamond by MAP Systems, Bengaluru, India

www.penguin.sg

My dearest Papa, thank you for always believing in me and pushing me to aim for the stars. All the morning assembly news reading and poetry recitations come handy when I find myself in rooms where I think I don't belong, I speak up because I know my voice matters. Even now I can hear you say, 'Nikku bete sab badia hoga' (Nikku, my child, everything will go well). Thank you for making me so comfortable in my skin that I don't seek outside validation.

To my greatest love, Momma, thank you for never giving up on me, and for owning my dreams as yours. You have always propelled my dreams by giving me wings to fly. I am what I am, thanks to your unwavering support and unconditional love. Thank you for making me believe in myself and God. With your blessings and love, nothing can ever go wrong.

Gratitude to Vanya, Aarav, and Nidhi for their constant loving annoyance; I almost sometimes forget that you are family, so I can't ignore you.

Contents

Why Did I Write This Book

Living in an era of rapid technological growth, we now find ourselves fully immersed in the vast frontier of Super Apps. While there is a lot of literature on the thinking and innovation behind Super Apps, there is no single resource that puts all of these methods together in a comprehensive and coherent way. This book covers stories of WeChat from China, Paytm and PhonePe from India, Kakao from South Korea, Zalo from Vietnam, KoinWorks and GoTo from Indonesia, and Grab from Singapore.

A Super App encompasses a wide array of services that are typically found in separate applications—financial payments and transactions, bill and utility payment, transportation services, communication services (social media platform), food deliveries, e-commerce services (mass marketplace) and more.

FinTech entrepreneurs, consultants, investors, and bankers are scrambling for information, this book can easily get them up to speed on the latest on Super Apps:

- Grasp the market dynamics of the 'Super App revolution'
- Realize Super App's potential and impact on financial inclusion and data privacy
- Gain founder's insight on their entrepreneurial journey

Super Apps have been garnering attention and success over recent years, particularly in Asian countries, where they are widely used.

The Western world will see this book as a wakeup call. The concept of Super Apps has not reached the Western world yet, which suggests that there is a long way to go for the development of the same, given that companies such as Google and Amazon-established in the West, have the capability to change the financial and economical scenarios in the world. The western developers and tech companies will be excited to re-shape their offerings after reading this book.

One Stop covers expert interviews sharing their perspectives and insights as to how Super Apps work, how they are revolutionizing the FinTech industry, democratizing access to finance and changing the way start-ups operate.

With the COVID-19 pandemic, the 'S' of Environment, Social and Governance (ESG)has taken centre stage. *One Stop* explores how Super Apps have emerged a winner in the financial inclusion space and ultimately create an inclusive and sustainable world for all.

Part I

Chapter 1

Time Travel

Once upon a time, believe it or not, there was neither a smartphone nor the internet. Could we ever in this current day and age imagine life without these?

Truly some of us have never known a world without the smarphone and internet. Every day, rather often several times a day, we use money in one form or another. Digital natives could never imagine the use of shell money for bride price ceremonies in the South Pacific. In fact, until a year ago during the pandemic, Port Moresby, and some more of the highlands of Papua New Guinea villages returned to the use of traditional shell money. The trajectory of money fascinates lawyers, economists, and technology experts. Dwelling on the history of money of ancient Greece and Lapita people of the Pacific, and its constant evolution with digital technologies, concerns each and every one of us.

The first mobile phone had simple but essential features such as a calendar, contact book, and calculator. The first-ever smartphone was released in 2002 by BlackBerry and it allowed wireless email after which the whole industry kickstarted mobile application development. Now, it has become a part of our lives as we use it for almost any task such as paying bills, watching movies, ordering food, browsing the internet, and chatting with family and friends. But let's try imagining an app that can conduct all sorts of services without having to leave the app at all or download each app for each service that you use. To assess all the services with one tap surely does sound like a dream, a dream that is nonetheless achievable. well, if we could make a device like a 'mobile phone', of course somewhere out there this dream of an all-in-one app exists.

This dream, however, has become a reality in the Asian markets. WeChat materialized this concept with their Super App invented in 2011. Yes, it's called a Super App, why you ask? Because it's an all-in-one app! Makes sense to call it a Super App, but WeChat was not always this way. Originally launched as a messaging app, it has now grown into an ecosystem of more than a million services. Its services range from doing trivial tasks such as booking a taxi to transacting with local businesses and government service providers. The programs are integrated into one app which gives convenience to the users.

How did Super Apps come into play? What led to the invention of a Super App and more importantly, what was life before Super Apps? Did the history of money lead to where we are today?

Single Click, Multiple Ticks: Our fast-paced lifestyle has made Super App a norm. But first, let's take a walk down history lane to explore the key developments and important events in the history of money.

Paper Money

During the Tang Dynasty in the seventh century, tradespeople in China began to leave their heavy amount of metal coins with a creditworthy agent, who would take note of how much money they had on deposit on a piece of paper. The paper could then be used to buy goods, and the seller could go to the agent and redeem the note for the coins and there it was—the invention of paper notes. In hindsight, this was completely accidental and people soon realized how feasible these paper notes were. China was able to use this system as they had invented printing facilities and paper. This simplified the storage of value and the process of trading. However, these promissory notes were still not considered true paper currency.

The concept of paper money constantly changed in the Song Dynasty in China. The government at the time issued the world's first national paper currency and they licensed various shops to allow an exchange of coins for paper notes, issuing its paper note currency.

Jiaozi was the first paper print money that was printed using woodblock printing and used six different colours and was valued according to the needs of the purchaser, which at first sounds good, doesn't it? However, it lacked standard denominations. The government was unable to regulate the production of lower denominations of the Jiaozi due to its immense popularity, which led to high rates of inflation.

The Yuan Dynasty created a cohesive, national system that was not supported by silver or gold. This was done by outlawing all transactions and possessions of silver or gold. To mitigate the issue of inflation, a new currency was developed by the government in 1287. Despite this, the inflation remained due to the same reason as Jiaozi: unregulated printing.

The Ming Dynasty attempted paper money again, however faced the same issue as the previous dynasties. Hence, the dynasty established a private system using the silver currency for all transactions. Was this the end of paper money?

Well, the answer is no. It was the Qing Dynasty, in 1889, that made a successful attempt at paper currency with the Chinese yuan.

The paper currency concept was introduced in Europe during the thirteenth century by travellers including Marco Polo who discussed it in his book *The Travels of Marco Polo* as well as William of Rubruck. Marco's visits to the court of Kubla Khan gave him an understanding of how the system of paper currency worked at the time. To quote Marco Polo: 'With these pieces of paper they can buy anything and pay for anything. And I can tell you that the papers that reckon as ten bezants do not weigh one'.[i]

Banknotes in Europe during the early days were issued by private banks as legal tender in the seventeenth century and this practice continued until the nineteenth century. The first attempt at banknotes was made in 1661 by a Swedish bank. These banknotes had a similar purpose to the paper notes in China to eventually convert them into gold or silver at the bank. Many private commercial banks in the United States were aggressively issuing banknotes. In fact, were close to 5,500 types of banknotes at one point. However, only the notes issued by the largest and most trustworthy banks were widely accepted and utilized. Eventually, the banknotes issued by private banks as legal tender were replaced with government-backed national banknotes. In 1913, the Federal Reserve Bank in the US was granted sole rights to issue banknotes.

Paper Credit Card

Though the advantages of using banknotes are plenty, it's not feasible enough, and has its fair share of disadvantages. Advancements in technology gave rise to digital money which was developed to mitigate the issues that arise from the usage of banknotes.

Counterfeiting of banknotes is the biggest peril of banknotes. Central banks around the world continue to take many measures to combat the counterfeiting of banknotes. Sadly enough, bad actors counterfeit and decide what quality to produce. India has carried out demonetization several times since independence to curb counterfeit currency notes of 500, 1,000, and 2,000. Counterfeiters in the neighbouring countries such as Bangladesh and Pakistan have been counterfeiting Indian rupees (a more widely used foreign currency rather than the domestic one).

Very ironically, the making of money costs a lot of money—since most banknotes are made with paper and are tangible, there are many costs involved. What are these costs? Well, primarily, the wear-and-tear cost and acceptance cost. Despite the condition, banknotes do not lose their intrinsic value. The poor condition of the notes means replacement with newer ones. To ensure the authenticity of notes, banks perform due diligence through multiple checks. These checks are part of what is called the cost of acceptance.

Fast forward to the nineteenth century when credit coins were first launched. Unlike banknotes charge coins were issued by departmental stores to extend holders' credit. The coins were given to people who had charge accounts in department stores, hotels, and more. The appearance of the coins was differentiated by their unique account number as well as the merchant's name and logo. Since these coins did not have the account owner's name on them, the chances of mistaken identity were higher. People could commit fraud on both the account owner and merchant by acting on behalf of the charge- account owner.

To combat this problem and make the system more feasible in 1958, the Bank of America created a revolving credit financial system similar to the credit cards of today. Since charge cards were only accepted by a handful of merchants based on which the merchant issued the cards, they were not as convenient. The system solved this issue by establishing a card that would be accepted by a wide variety of merchants. The paper card was named BankAmericard, with a $300 limit. This credit card offered revolving credit and was the first step towards a modern-day credit card. The BankAmericard gave people the ability to carry a balance and proved to be a success in convincing both the merchants and consumers to use the card. The card was then licensed to other banks and nations. In 1976, all licensees united under the common brand called 'Visa'. To compete with BankAmericard, MasterCard was created by a group of banks. Credit cards became useful for those who were travelling around the country to shift their credit where they could not use their banking facilities directly.

And then from here on, the rise of digital money started, a new era, a new concept that disrupted the financial industry and economy.

The Rise of Digital Money

With the emergence of the internet in 1990, online payments became more preferred among consumers. The first idea we will explore will be a digital currency. But what even is a digital currency like currency but digital? Well, it is a currency that is mostly managed, stored, or exchanged on digital computer systems, such as the internet. This lack of tangible aspect that a typical credit card would provide allows instantaneous transactions over the internet and removes the aforementioned costs with the banknotes. The first mention of the idea of digital cash was in a research paper by David Chaum who later founded DigiCash in 1989. The first digital currency that was popularized by both consumers and merchants in the 1990s was e-gold, a digital gold currency. Users could make instant transactions to other e-gold accounts. In 1998, PayPal was launched as a payment method with USD as the underlying currency. The users were no longer required to share credit card or bank account details. PayPal used VeriSign as well as acquired Fraud Sciences (a privately held Israeli start-up) to enhance its risk tools. PayPal strengthened its payment mechanism to ensure secure and seamless transactions, making it one of the most valued start-ups and widely used payment methods in the world. The PayPal mafia (founders, and the initial team) went on to launch some of the most successful companies in history namely Tesla, Inc., SpaceX, Affirm, Slide, Kiva, LinkedIn, Palantir Technologies, YouTube, Yelp, and Yammer. Elon Musk and Peter Thiel (the don of the PayPal mafia) need no introduction in the business innovation world.

Okay, so we have digital money but where do we even use this? Why use digital money in physical shops with tangible goods? Is the digital market better equipped for digital money? Another technological innovation that experienced immense growth was e-commerce, the buying or selling of products online or over the internet. The early e-commerce sites included eBay, Alibaba.com, and Amazon.com. The first business-to-business online shopping system was established by Thomson Holidays UK in 1981. In 1992, a commercial sales website using credit card processing was established by Book Stacks Unlimited.

The first online auction site, eBay was founded in 1995 for person-to-person transactions. The payments were supported by PayPal allowing eBay users to pay through their PayPal accounts. PayPal became extremely popular in the early 2000s, more than 70 per cent of the auction listings on eBay were settled via PayPal. In 2001, eBay became the biggest e-commerce site and in 2002, eBay went on to acquire PayPal.

Another notable development is the advent of internet banking. It is a system that allows users of a bank or other financial institution to conduct a range of transactions through the website of the bank. This provides users access to banking services instead of going down to the branch to make that same transaction. Internet banking reduces banks' operating expenses while providing convenience to customers, allowing them to make quick transactions on the internet rather than going about the inefficient method of branch banking.

Digital money along with the growth of FinTech (Finance and Technology) led us to the Super App story which we will look at in the next few chapters. Paper money alone would not have led us to Super Apps but all these developments jointly played a crucial role to help us reach where we are now: One Stop. Super Apps.

2008 Financial Crisis

The 2008 financial crisis occurred due to lax financial regulation, excessive risk-taking by banks, and reduced valuations of mortgage-backed securities which were tied to American real estate. There was also the increased use of leverage in the financial system- banks extending loans to borrowers who were previously considered untrustworthy. As the approvals went up, demand spiked up, increasing the housing prices as a result. Since in the first place, loans were extended to not-so-creditworthy borrowers, the default rates went north, leading to a rapid devaluation of mortgage-backed securities. This then sparked the Great Recession that contributed to the increase in unemployment and reduced institutional trust.

One of the main causes of the financial crisis was the subprime mortgage crisis. This crisis began when there was a sharp decline in home prices in the United States after the collapse of a housing bubble. This ultimately led to foreclosure, reduced values of mortgage-backed securities as well as increased mortgage delinquencies. There was a huge impact on the investment in housing which was then followed by reduced household spending and business investment. The houses that led up to the crisis were supported by mortgage-backed securities that offered higher interest rates than the national securities and good risk ratings from rating agencies. These attractive benefits were what encouraged potential homebuyers to take up the loan and led to the increase in subprime lending. A high fraction of these subprime mortgages was tied to a floating ever-growing interest rate. The motive of using this was for customers that are not able to pay large amounts of initial payment. Customers must accept the risk of interest rate changes for a smaller initial payment. This was a part of the trend of reducing lending standards and selling higher-risk mortgage products such as the one mentioned above. This, however, backfired as many were not able to afford the payments of the mortgage, leading to more and more households in debt. Other key developments in this crisis include mortgage fraud and predatory lending.

Financial Crisis: The Birth of Fintech

The subprime mortgage crisis collapsed the global economy, Lehman Brothers went bankrupt and banks were bailed out with taxpayer's money. Needless to say, the public lost trust in the banking modus operandi and the world of finance in general. This made for a fertile ground for FinTech to take birth and lay the foundations to shake up banks' long-existing monopoly. For the longest time, banks faced little to no competition, but now this was set to change with innovation opening doors for transparency in fees, offerings, and exchange spreads (FinTech).

Many bankers ended up being disillusioned by the financial meltdown and decided to turn towards entrepreneurship. Companies such as Funding Circle, Zopa, Crowdcube, and GoCardless, and a slew of digital challenger banks, such as Revolut, Starling, and Monzo came about. Paving way for frictionless payment, lending, and remittance- bringing down the cost of sending, spending, lending, and receiving money (FinTech). In no time some of them became the fastest-growing fintech unicorns.

The collapse of Lehman Brothers in 2008 forced banks to look inward to avert future meltdowns. FinTechs on the other hand went down the technology route and embraced themselves to be agile, cost-effective, and work on customer-centric models.

This was also made possible for the consumer trust in upcoming fintech over the traditional banks.

FinTech: Continue to Eat Banks Lunch

It is not difficult to connect the dots from the seeds of distrust to the starting of a new era, Fintech. Twelve years later, FinTech and Super Apps continue to eat banks' lunch and dominate front-page news headlines. What does the future hold for FinTechs? By the end of 2022, the Fintech market value is likely to cross a valuation of $309.98 billion. A consistent uptick in investment rounds displays a strong potential for FinTechs. In 2008, the total number of Fintech investments was just under $1 billion, and that number as of 2021 is $210 billion.

The use of emerging technology, huge venture funding rounds, and a strong commitment to serving the last mile makes a good use case for Fintech. Similar to the 2008 crisis, the coronavirus pandemic accelerated the penetration of FinTechs. It has helped in bringing *the offline to online switch* for the wider population. While the financial crisis fundamentally altered our relationship with banking and birthed FinTech. COVID-19 birthed innovation to the existing Fintech models to cater to the new normal-telemedicine, ease of going online, contact tracing, dispersing information on vaccination and lockdowns, and staying connected with our loved ones. This period put a spotlight on sustainability, SDGs to protect our climate and bridge the digital, financial, and gender divide through FinTech and Super Apps.

COVID-19 and its Impact on the Economy

The economic contraction since 2020 due to the COVID-19 pandemic is the worst since 1947 and the COVID-19 stricken economy is often compared with the economy during the Great Depression. In June 2020, the International Monetary Fund (IMF) reported that the expected global growth to recess by 4.9 per cent. Employment has been severely impacted where some countries had more job cuts than new job creation which resulted in a high unemployment rate. The pandemic has provided a breeding ground where it is more likely to have a significant increase in inequality and poverty. Sectors such as air transport and tourism have been severely affected due to COVID-19 regulations and restrictions.

The COVID-19 pandemic has impacted the expansion plans of Super Apps. As a result, companies now focus on consolidating their products through cutting services that bring less profit compared to others. Grab's core project was its ride-hailing services but the rise in work-from-home reduced the need for ride-hailing services. In 2020, Grab—a Singapore based super-app—laid off around 360 employees after its ride-hailing rock bottom including its biggest market Indonesia, implemented lockdown measures. Now the food business yields more[ii] than 50 per cent of the revenue, owing to increased demand for food delivery. As COVID-19 kept more people indoors, food delivery saw a rise in usage. Hence, Super Apps like Grab from Singapore pivoted to focus more on services such as insurance and GrabFood and GrabMart, which are its food and grocery delivery services.

Financial services, in addition to food delivery, play an increasingly crucial role in the future plans of Super Apps. Following the introduction of GoPay and GrabPay to process payments for ride-hailing bookings, both Gojek (now GoTo) from Indonesia and Grab rapidly expanded into financial services by including point-of-sale and online payments, as well as ride insurance for its riders and drivers, travel insurance, and small and medium-sized business loans. In the first quarter of 2020, Ming Maa, President of Grab, shared that 'access to services such as cheap health care, insurance, smart mobility, and financial services remains a critical hurdle [to growth] in the region.' 'These are difficulties that multi-service digital platforms must address as they extend their offerings to provide users with more ease and choice'[iii]. Given the rise in e-commerce that has accompanied the epidemic, the industry leaders in Southeast Asia are aggressively expanding their financial footprints even further. In April 2020, Gojek completed the $130 million acquisition of local Indonesian payments start-up Moka, and in June, Facebook and PayPal announced their participation in Gojek's Series F round of funding, which raised roughly $3 billion.

What is a Super App?

Having mentioned 'Super App' a number of times already it's about time to unpack it. Simply put, a Super App is a one-stop-shop for a multitude of services, including both in-house and third-party offerings. First incorporated in China but soon travelled far and wide across borders and made its way into the global economy. The most prominent names in the Chinese and overseas markets still remain WeChat, and Alipay besides Grab, GoTo, Paytm, and Kakao Talk.

What Makes Super Apps Super?

Using a slew of apps and remembering their passwords is no easy feat. Super App solves that problem by being an umbrella of a diversified portfolio of services, where one password is all, you need. In return, the app captures a larger pie of user wallets and time. Sometimes, the aggregation of services is not-so-subtle and puts off the users. The rat race to become the next Super App is a dream harbored by many including the tech giants Google, Facebook, and Amazon.

Super App-Superheroes for Small Business

TenCent, the Chinese tech giant's baby venture WeChat is a leader in the Super App space. WeChat started off as a messaging app soon morphing into an ecosystem of services: taxi rides, medical consultations, payments or virtual wallets, hotel reservations, games. With a billion-plus users under its kitty and the fifth most popular social networking site worldwide, WeChat does not need to market the apps to a whole new audience, every time something new is launched and integrated into WeChat, it is automatically available to its entire user base. They are also able to choose the right products to launch in the app as they have data from users that can show the usage of the programs that it offers. The data about users' preferences can be analyzed to infer what products should be integrated next. As Super Apps can integrate products or mini-apps by purchasing the license to run the program or collabourating with companies that want a piece of the huge audience that it can market to. This allows companies to have a more cost-effective strategy when releasing a product—expanding their range of services whilst capturing a wide base of Super App audiences. The users also feel confident and safe in using the offerings as it lies within the app itself, a space with which they are familiar.

Now that we have taken a peek into how Super Apps have come about, let's move on to find out how different parts of the world have perceived Super Apps, mainly the East and the West!

Chapter 2

One World, Two Cultures

Before we dive into how exactly Super Apps work, let's see how Super Apps have affected different parts of the world, differently. We will be focusing on the dynamic East-West economic relations that have been central to multiple debates and discussions, mostly at global levels, and how they influence capitalism, mainly China and America. To start, have you ever wondered what the Asian Age is?

The Coming of the Asian Age

The *Lord of the Rings* metaphor for Super Apps is not off track— one ring that rules them all; one app that rules them all—as it is a platform that combines multi-purpose services into a single experience. This one-tap app goes beyond providing services as it also provides unique personalized experiences. China enjoys the status of a leading global force for Super Apps.

With technology evolving rapidly, we have now entered the digital rat race, once upon a time the IT sector was dominated by Americans, well that's not the case any more the whole globe is collectively moving towards digitalization! Well, maybe not the whole globe but, a majority of it is in the East. But a lot has changed over the years as now the tech-savvy millennial and GenZ have taken over the global market.

In Asia, tech platforms have turned financial services into digital utilities. They act as platforms that allow for networking, transportation, and even e-commerce. In Asia, there is excessive usage of these digital utilities like Alipay, WeChat, GoTo and Grab which act as Super Apps, combining a plethora of services under one platform. It is convenient, efficient and its multi-functionality continues to grow at a fast pace in Asia, giving tough competition to the West.

China and India have been setting trends, with the latter posed to become the world's third-largest economy in 2030, you know what this means– it's another hot spot for Super Apps, India is all set to redefine its mobile landscape. Exports have surged, the growth to GDP ratio is high, and foreign direct investment showcases an upward trend. In 2019, SoftBank had agreed to invest in Paytm's latest $1 billion funding round on the condition that it goes public within five years of completion of funding. (Paytm's parent company One97 Communications raised $1.4 billion in funding from SoftBank Group, the Japanese internet and telecom major. SoftBank also joined Alibaba Group as a major shareholder and in turn, got a seat on the Paytm board). The much-awaited IPO at the Bombay Stock Exchange (BSE) opened without much fanfare, the stock prices tanked on its opening day. The downward spiral trend continues, infamously earning the company tag of a 'cash guzzler'.

Half of the world's population has no access to the internet (4.66 billion people in the world have access to the internet). Fewer than one in five people in the least developed countries are without internet. The advent of internet technology has opened new doors of limitless possibilities far more for the East than for the West. Almost 2.9 billion people or 37 per cent of the world's population–have still never used the internet whereas[iv] as of January 2023, China ranked first among the countries with the most internet users worldwide. World's most populated country had 1.05 billion internet users, more than triple the third-ranked United States with just around 311 million internet users.[v] China's educated, tech-savvy millennial group and a large number of investments in start-ups; all of these factors have made Asia much more desirable than the US. Many political scientists and economists have deliberated on the expectation that the Asian economies will be the next big thing. The 'Asian Age' is the tailwind for economic growth in the FinTech age.

The East Versus The West

Culture impacts the business environment. Understanding cultural nuances, barriers and diversity not only prevents pitfalls but also aids in finding a good market fit.

So, what makes the East and the West so different?

Here's the Western culture for you:

- In the Western world, Judaism, Christianity, and Islam are some of the most widely practiced religions.
- Westerners are more likely to have liberal values and often have open spaces to discuss taboo topics. In the Asian culture, however, topics like sex and the experience of childbirth are often hush-hush.
- People in the West uninhibitedly express their emotions. Asian culture is sometimes synonymous with people-pleasing and trying to fit in rather than questioning the status quo.
- More often than not, Asians are crippled with guilt for putting their personal life over family responsibilities. In the West, people take gap years for soul-searching or exploring the world. From a young age, they learn to put themselves first and are more likely to pursue their happiness.

Cultural and educational contrasts may exist between the East and the West. The majority of these distinctions can be seen in people's conduct and attitudes.

In the Eastern Culture:

- In the Eastern globe, Hinduism, Buddhism, Jainism, Shenism, Taoism, and Islam are some of the most widely practiced religions.
- When it comes to clothes, customs, and other aspects of life, the inhabitants of Eastern countries are more traditional than those of Western countries. For instance, in Asia, as a sign of respect or acknowledgment, people often bow or greet with a Namaste by joining their hands.
- People in the East hold tight to their views and beliefs. They don't tend to question rituals and traditions.
- In Eastern countries, the elders are regarded as the home's leaders and often take important decisions, without consulting young members of their families. The concept of personal space is almost non existential.

Now that we have looked at some of the general practices, how about some business practices generally followed in these two worlds?

1. Relationships

When it comes to business partnerships, it can be difficult to know what to expect when you travel to a new country. Understanding that expectations may differ from what you're used to is the first step towards success. People in the West are more likely to prefer formal meetings for conducting business. When you're at work, you're at work, and it's normal for employees to avoid forming close bonds with one another. The occasional office get-together, on the other hand, provides an opportunity to bond through any mutual embarrassment that may occur.

When it comes to employee-employer interactions, Western countries have lower standards for how well firms look after their employees (beyond what is required by law). When Americans get laid off, employers give them just hours to pack their belongings and leave. In Japan, layoffs are even considered a social taboo.

The majority of Westerners are unlikely to be startled if their contract is terminated due to poor performance.

Employees, likewise, have no qualms about leaving their current job for brighter pastures if the opportunity arises.

Personal sharing and the formation of closer long-term friendships are favoured and promoted in the East, rather than keeping relationships strictly professional. There is a desire to make connections with co-workers, which takes time and trust to develop, and people can be angered or ashamed if others do not reciprocate. Efforts to strengthen these bonds are typical. Activities like karaoke, for example, are a fantastic illustration of after-work contact between colleagues in Japan.

In comparison to the West, corporate Asia is less stringent when it comes to hiring and is more indulgent with under-performers on staff. Employees have a strong attachment to their employers, and there is a cultural expectation that companies would look after them. People dislike working with strangers and find it difficult to develop relationships. However, once a relationship is established, it is common for it to last a lifetime.

2. Criticism

Depending on the work culture you're in, what constitutes criticism—or even where the line between guidance, expert opinion, or seniority is drawn—can be very different.

In the West, it's quite common to call individuals out on their faults. In fact, it's frequently seen as an essential component in building a strong and productive team. It's perfectly fine to confront difficulties directly and inform co-workers of their shortcomings or faults, albeit the manner in which you do so may differ by territory. In the United Kingdom, making a snarky joke about a problem may be useful, but it may be disrespectful in Germany, where people prefer to receive only the facts about their mistake and nothing else. In the United States, a courteous email would suffice. Whatever technique is taken, the reality remains that in the West, problems are identified and blame is placed on those responsible. Additionally, the associated rage, irritation, and other relevant emotions may be displayed.

Criticizing a co-worker in front of the rest of the team is nearly inconceivable in the East. People will strive to stay as far away from these situations as possible. In fact, if there's a method to avoid unpleasant situations entirely, this is the best path to follow. For example, in China, the concept of 'preserving face' is ingrained in society. Criticism is generally conveyed through a third party and is kept for private contacts. The face is a difficult concept to grasp, and if you're planning a trip to China, we recommend doing extensive research on the issue.

People in the East are generally quieter and don't readily express their emotions. Harmony and the avoidance of conflict are regarded as important goals to pursue.

3. Questions

The act of just asking a question can be perceived in a variety of ways by both the person being asked and by observers. When interning in Eastern versus Western cultures, you'll rapidly notice these distinctions.

In the West, asking questions is considered a routine procedure; in fact, it is expected that lower-level employees take the initiative to learn more about important topics by asking questions about them. Employees are encouraged to raise questions, even if they are critical of their bosses' views and conduct. It's not so much that you're disputing ideas as an employee as it is that you're simply trying to get your head around them and better understand motives and processes. Employers encourage this type of behaviour because it demonstrates that the person asking the questions is motivated to learn more in order to become a more valuable employee. After all, leaders in the West are just another member of the team.

Things function differently in the East. The thought of asking inquiries would most likely terrify employees. There is a risk that supervisors would perceive questioning as threatening since they will have to clarify their viewpoint on a certain topic. The necessity of civility is emphasized far more than the importance of openly sharing thoughts, disputes, or pointing out mistakes. The East's leadership is precisely that, and it should not be questioned.

4. Authority

In any workplace, you will come across hierarchies. The harshness of that status and authority system, on the other hand, may vary dramatically between regions and territories.

Hierarchies in the West tend to be quite flat. You should be able to talk to your boss, and even the company's CEO, without hesitation. You might even be able to address them by their first name. People in higher positions will usually try to make their subordinates feel like they are on an equal footing with the rest of the team. Every individual's viewpoint is valid, and all ideas are welcome. Final decisions are made based on the input of the team, and no one agrees to something just to agree—there is a shared belief that this is not how problems are handled. Regardless of the standing of the individual expressing the concern, potential issues should be recognized as soon as feasible and reconsidered. Furthermore, it is normal for employees to criticize one another in front of their co-workers. Because you're only sifting through ideas, it's generally good to turn off distracting input.

In the East, the hierarchical system comprises many more layers, each of which has a distinct meaning. The highest-ranking officials have the last say, and the system of order and government is seen as crucial. It is true that seniority matters. In the Eastern world, boss directives means everything, while the Western world fosters equity and informal structures (vI). Interns in Singapore, for example, frequently observe that Chinese ideals of *guanxi* (relationships) and hierarchies are prevalent, publicly criticizing superiors and defying orders is a big NO.

It all comes down to position and authority, from how each individual is greeted (and in what order), to how they are served drinks (for example), and how they are interacted with. Managers will rarely reprimand their employees or associates in front of others, and any advice or input will almost certainly be provided with subtle inferences and non-verbal indications specific to that culture or location. Bosses are sometimes revered as father figures, a practice that harkens back to the long-term ties and strong bonding we described before.

However, it is important to take note that these characteristics are generalized and may not be true for each individual or business. Cultural understanding helps to define nations and its people— their spending habits and behavioural pattern. Businesses that factor in cultural aspects connect better with their customers and employees in turn achieving greater efficiency and predictability.[vi]

Capitalism and Its Influence

It is 2022. Russia has invaded Ukraine. Most agree might is not right. Many believe that the Russian political capitalists waged the war to survive, thrive and to continue creating wealth by stealing the state. Capitalism influences our daily lives, how countries, businesses, and companies think and structure their plans. Even future moves are all based on the concept of capitalism, let's look at capitalism and its influence.

In a capitalist world, supply and demand factors decide the price rather than a central authority. Private individuals and organizations with their know-how, capital, raw materials, and production tools manage their own production output. In a laissez-faire market or free market, there are no checks or regulations-private individuals make their own investments decisions, chase their entrepreneurial callings, create unique products, and decide the rates at which commodities and services are exchanged. This is the purest form of capitalism with the unrestricted flow of goods and services, and ideas are the currency of the twenty-first century. The basic element of capitalism is the right to own private property and as long as the owner keeps within the confines of the law, which in capitalist systems are often broad, the individual is free to do whatever they want with their property. A peer-to-peer transaction for the sale/purchase of property or any asset is done at a mutually agreed-upon price rather than a decision made by the government. The prices at which property is bought and sold in a capitalist society are determined by the free market forces of supply and demand, rather than by a central ruling authority. The foundation of capitalist production is private property rights. These rights clearly distinguish between those who control the means of production and those who use them. An entrepreneur, for example, will own both the factory and the machines that run it, as well as the finished product. A worker who works within that factory and uses those machinery does not own them, and they cannot take the finished product home with them for personal use or sale; that would be stealing. In exchange for their labour, the worker is simply entitled to their wages.

Private entities by virtue of entrepreneurship, land, capital, and labour produce goods and services. Such companies use a combination of these elements to maximize profit and efficiency. What happens to surplus products is a classic measure of whether the factors of production are privately or publicly held. Surplus product is provided to society as a whole in a communist economy, whereas the surplus product is held by the producer and used to generate extra profit in a capitalist system.

The actors or players of a capitalist economy run on ambitions and a strong desire for profit. Financial motivation pushes people to work hard, filling market gaps by offering new products or services. The intent is to increase net wealth by creating, innovating, and ideally beat the competition by being the first mover in the market. As Marxism shows, capital re-deployment is required for increased market share by improving productivity and efficiency.

Today's Tinder has its roots in capitalism—the right swipe for a good-looking profile. The free market can be compared to the animal kingdom—rivalry and survival of the fittest. A brand survives if it's better and less expensive than its competitors. For instance, a Chinese product in an overseas market tends to be sold like hotcakes since it's unique and cheaper than homegrown products in that market. Businesses continue to look for ways to maximize efficiency and provide competitive prices to be the market leader and rule consumers' hearts. After all, one could be more loyal to a brand than to a Tinder date. In a capitalist system, businesses decide who to do business with; however, in a communist system, the central authority tends to establish monopolies in all industries.

In real life, capitalism and socialism and communism tend to co-exist. In a mixed economy, a hybrid model of the private-sector system of capitalism exists along with the public sector business system of socialism. The government intervenes to inhibit monopolistic behaviour to avoid the concentration of economic power. In such a system both the state and people own resources.

Delving into Power Players of the East and West: China and the US

Though there are a lot of cultural differences between the East and the West, people from these distinct societies also have just as many if not more personal commonalities than differences.

Comfort, money, providing for the family, job fulfilment, and security are all motivators for people. Nevertheless, each culture achieves these aims in various ways, and on the surface, they may appear to be poles apart. Understanding why individuals behave the way they do can help to communicate more effectively across cultures. The 'American Dream' has drawn many immigrants to the US. In contrast to this, in Chinese culture the group is very important and individualism takes a backseat. Any achievement is credited to the family or the team. Hierarchy in the organization is given due respect in China, the junior staff seldom interacts with the top management. American companies do away with hierarchy sometimes and promote an open culture.

Asians especially Chinese and Indians deeply respect their elderly folks and sometimes continue to live in the same house with them after marriage. Multiple generations often live in the same house sharing household chores and mostly taking care of their elderly. They have traditions to pray for their deceased ones. Americans however, expect their kids to be independent and leave the house once they attain the legal age of eighteen. Senior care homes are widely popular in the States.

Guanxi (relationships) is a way of life for the Chinese. Americans on the other hand are known to keep their work and personal life separate. *Guanxi* extends to business too, building relationships with co-workers is seen as necessary—after-work parties and drinks with colleagues are part of the Chinese work culture.

The pace of business in the United States differs from that in China. Americans place a premium on speed and efficiency, and they will rush to complete tasks. Meetings are expected to start on time, and deadlines are expected to be met. The Chinese, on the other hand, are known for taking their time making decisions, preferring to develop an agreement and cultivate relationships before committing to anything. Only when the moment is appropriate and the project is regarded as complete can deadlines be met. Americans may find this timeliness attitude irritating and time-consuming, while the Chinese will take advantage of the Americans' demand for speed in negotiations, playing a waiting game in order to gain a better deal for themselves.

How Do these Differences Affect Technological Innovation and the Economy?

A new 'Great Game' between the United States and China is centred on technology, and it is a game that counts. The focus of US trade and investment policies has shifted from lowering the bilateral trade deficit to technology. There is a geopolitical component to this, but economics also plays a role. The United States' criticisms centre on sectors where technology is dominant, such as intellectual property protection, access to fast-growing Chinese markets for American enterprises, and a level playing field against domestic Chinese champions. Friction is being felt throughout trade and investment partnerships as the world adjusts to a growing China. Technology—how it is traded, transferred, and regulated—is one common thread underlying these frictions. The stakes for China are rising as the focus shifts from trade to technology. The rapid and efficient deployment of new technologies across China's economy is possibly more important than anything else in the country's rebalancing and catching-up to rich countries. Being the most populated country in the world, needless to say China has access to a huge labour pool. However, now the country is turning its focus on FinTech and is often hailed as the big tech giant in the world. It is known for its supportive regulatory policies, strong mobile infrastructure, thriving start-up ecosystem, or the ban on cryptos and its mining- a move which is intended to promote its own digital currency for example, Digital Yuan(e-CNY) also known as the Central Bank Digital Currency (CBDC). People's

Bank of China (PBOC) is making every possible effort to be the honey pot of data by replacing the most commonly used payment methods (privately held)—the Super Apps such as WeChat and AliPay, and driving traffic to e-CNY (owned and has sovereign backing by the PBOC).

While central banks around the world are struggling to get their narrative right on CBDC, China has implemented it successfully. China has Offered more than 50,000 Shenzhen citizens digital Yuan of $30 in a lottery—two million applied to participate, and giving away digital Yuan in red packets (a gift given in the Chinese tradition) on the Chinese New Year, created the right spur[vii]. The citizens can use it at shops across the country. Adding to that momentum is the fact that there are no private banks in China—all banks have some extent of state ownership leading to no lobbying pressure.

Let's not forget this is part of a mighty plan, an attempt to overtake USD (de-dollarization). By carrying out bilateral trades in the Digital Yuan or Yuan, China is making a small yet strong move that would aid in replacing the USD domination in international trade and reducing the current foreign reserves in the USD. With the increasing list of sanctions on Russia, Digital Yuan or Yuan is expected to gain further momentum.

Digital assets such as bitcoin are increasingly used for inflation hedging amidst the negative interest rates pushed by the central banks around the world. European Central Bank (ECB) and US Federal implemented such rates to push consumer spending and support the local economy post covid crisis. Consequently, leading to a positive mind-shift towards digital currency. Bitcoin is hailed as digital gold and people are taking solace in it for asset hedging. This coupled with the Russian attack on Ukraine and European Union banning the SWIFT channels for Russia, it'd be interesting to see how the Euro fares which in turn might lead to strengthening Yuan's global standing.

The direct demand blow to China's economy caused by friction with the US is likely to be tolerable in the short term. The tariff hit is minor because U.S. exports account for less than 4 per cent of China's GDP, a figure that drops much lower when measured in terms of added value (netting out imported inputs). It's difficult to sway China's macroeconomic policy on bilateral trade with the United States, and any repercussions might be mitigated by a combination of moderate currency depreciation, trade diversion, and fiscal stimulus.

Now that we have looked at the differences between the East and the West, specifically China and the US, let's look at how these characteristics that we have seen have resulted in the West being hesitant about Super Apps!

With this hesitation of the West stopping itself from tapping into Super Apps, is it safe to say that the 'Asian Age' has the potential to change the tone of the economic world?

Chapter 3

Hesitancy

No doubt Super Apps are a big hit in the East; however, they are not gaining traction in the West as it has got in the East. Is there still a possibility for Super Apps to thrive in the Western market? There are many factors stifling the Western world. This can range from fundamental reasons such as the strong importance of privacy to external factors such as controversies that have shaken the technology industry. Even though the future for Super Apps in the West seems unclear, that does not stop companies from trying to launch one. Revolut and Uber have been taking strides in becoming Super App and their steps are not small either. Let's look in detail into the possible reasons for the West's hesitancy in regard to Super Apps.

Possible Reasons for Hesitancy

What reasons could there be that Westerners have not tapped into the era of Super Apps yet? And if at all they ever do decide, what key factors would they take into consideration? Westerners are more likely to explore several options before purchasing something, which can be related to the characteristics we have discussed previously. Hence, a sophisticated balance between practicality and flexibility, such as how useful and simple it is to use the app, needs to be developed to entice the Westerners. Trust, which is a foundation of relationships as discussed before, also needs to be built to tackle privacy concerns due to the controversies caused by Facebook and Google. Westerners may not be so willing to give away privacy for convenience. With the high amount of granular data collected, which is detailed data, or the data in a target set at its most basic level, companies need to provide a high level of guarantee with the implementation of safeguards. There are also regulatory constraints in the US and Europe, which have become tougher. This has prevented the emergence of a provider capable of handling most of our daily needs. In addition, the competition law also prevents the mergers of big tech players together with the data protection laws to aggregate personal data in Europe.

Desktop to Mobile Experience

When iPhone launched its App Store, the top early downloads at the time included Facebook, Google Mobile, eBay Mobile, and more. This was what defined the pre-existing internet environment and customers got accustomed to it. As people moved from the browser to phones, these primarily desktop browser users expected phones to be a continuation of that experience.

Competition just to dominate in one space was incredibly tight, and companies put their best efforts to succeed where they had captured one area of the customers' attention. By the time companies like Google, Facebook, and Apple had accumulated enough cash and runway to be secure in search and ads, social media, and consumer electronics respectively, there were already other companies blazing the way for shared economy, health, finances, and all else.

Controversy in the West

One of the biggest controversies in the West on the technological side would be the data protection laws that have entered into force. In 2016, a package of data protection measures was undertaken by the European Union (EU). This includes the General Data Protection Regulation (GDPR), a board of European authorities known as the European Data Protection Board (EDPB) that mandates European Data Protection Supervisor (EDPS) as well as appoint a Data Protection Officer to monitor and apply the rules.

The GDPR protects the public's personal data by regulating and monitoring the processing and movement of data. This law does not only affect firms operating in the EU but also other organizations around the world that offer products to or monitor EU citizens. This regulation is daunting for most organizations as the compliance costs for this are extremely high which can go up to €20 million or 4 per cent of global revenue, depending on which is higher. Firms are required to process data in a lawful and fair manner. This means that processing or the use of data should be within the stipulations agreed upon. There is an added responsibility in ensuring that the data is accurate and that the data is kept for a reasonable duration based on the purpose it is used for. Measures should be taken to ensure the security of data such as encryption, two-factor authentication, and staff training on the handling of sensitive data. In the sections below, we will take a look at two examples of data controversies that have shaken the West to better understand why they could possibly be hesitant about adopting Super Apps.

Facebook vs GDPR

One example of controversy would be Facebook's non-compliance with GDPR in October 2018. Facebook was close to being fined $1.6 Billion for its role in the Cambridge Analytica controversy. In 2010's, personal data of 87 million Facebook users was collected without consent by the UK-based political consulting firm, Cambridge Analytica, to help their clients expand their potential voter base.[viii] Ultimately Facebook was fined

£500,000 by the UK's data protection authorities for its repeated data breaches. One of the data breaches allowed hackers to infiltrate users' accounts by taking advantage of the vulnerability in one of its features and an estimated number of 50 million users were affected by this breach. While Facebook notified users and authorities of the breach within the time limit set by one of the regulations, it left out important details.

While Facebook has made the effort to comply with GDPR, it has left out some vital requirements in the GDPR. One particular effort they have made to comply with the regulation would be the transparency of data stored by allowing users to download an extract of their data stored on Facebook. Another company that gives the ability to download the extract would be Google. Facebook also offers options in the settings to give users control over the data shared with Facebook's partners. However, it is not a foolproof way to have complete control over your data. There are two main issues in its measures of complying with the GDPR. Facebook is a platform to connect with others but users are not able to control the use of their data if the data is shared with others. However, this does not comply with the GDPR as the consent of sharing data is ambiguous and reliant on the action of sharing the information. Not many users may realize that sharing of the information may be subjected to the use of data as it is not explicitly stated. Another issue would be the lack of coverage of teens' privacy. Countries in the EU have the regulation for parental consent, if the children are under the age of sixteen.

Google vs CNIL

In January 2019, Google was imposed a financial penalty of €50 million euros by the CNIL's committee due to its non-compliance with GDPR. The National Data Protection Commission (CNIL) in France received several group complaints. The complaint accused Google of the lack of a valid legal basis to process the personal data of the users, especially for ad personalization purposes- None of your Business ('NYOB') and La Quadrature Du Net ('LQDN'). The authority used the European framework that is recognized and utilized by all European authorities in the EDPB's guidelines. The two main violations made by Google were the lack of transparency failing which users couldn't truly understand the extent of the processing of information and the lack of legal basis for ad personalization processing.

The data processing purposes, data storage periods, and categories of personal data used for ad personalization weren't clearly laid out by Google. The information is not easy to find as it is dispersed across a number of documents and users have to click on links and buttons to access additional information. The extent of how massive and intrusive the processing operations of Google is not often understood by the users. The purpose of the processing as well as the categories of personal data is often vague, which is not clear enough for the users to understand the legal basis of the processing of data for ad personalization.

The CNIL also considers the user's consent as not valid. As the users do not realize how their data is processed, hence the consent that they give cannot be valid. According to GDPR, the user's consent must be unambiguous and specific. Although the users are able to modify their preferences on how their data can be used or collected, it takes additional steps for the users to change their preferences. Moreover, the option for display of ad personalization is pre-ticked which means the consent is unambiguous, where it is assumed that every user agrees for their data to be used for ad personalization. It requires users' actions to untick the option. However, not many are aware of this option. Many would assume that their data is not being used for ad personalization since it was not explicitly asked. This could cause users' data to be used without their explicit consent. The consent can only be considered to be specific if the consent is given for each purpose which is not the case with Google. When a user creates an account with Google, users are shown the terms and conditions and the option to agree to Google's Terms of Service, and the processing of information is described in the Privacy Policy.

Eventually, people would realize mistakes are for learning not for repeating. Google went on to repeat the same mistake. On 31 December 2021, CNIL fined Google €150 million, this was based on many complaints about the cookies on the websites google.fr and youtube.com. Google LLC was fined €90 million and Google Ireland Limited €60 million. The CNIL stated that the refusal mechanism (several clicks are required to refuse all cookies, against a single one to accept them) makes the refusal more complex in turn discouraging users from refusing cookies and encouraging them to opt for the ease of the 'I accept' button.

How Does the Controversy of the West Affect the Creation of Super Apps in the West?

Technology firms like Facebook face compliance issues due to a lack of consent and control due to their design. This is a similar problem faced by Super Apps such as WeChat. For the sake of easier understanding, we will use WeChat's app framework as a possible adoption by technology firms in the West. Tencent's WeChat has been alleged to be handing over user data to authorities in China without legal obligation to do so. There is a lack of control and consent in how the users' data is generated, collected, processed, and shared. There are no boundaries as to how much data a Super App can collect on users which can risk compliance issues as data may have purposes outside of the agreement between the firm and users. This is because a large amount of data is generated from a Super App, which allows the companies to deliver better customer experiences and product recommendations. Hence, it is important for Super Apps to determine the parameters by which the data will be collected and processed. It is also important to take note of the security measures that Super Apps undertake for its data stored. As security breaches and threats are on the rise, measures must be taken to mitigate emerging risks from the vulnerability of the Super Apps. Controversies like Facebook vs GDPR make it harder for users to trust firms with their data. Hence, it will be harder for Super App companies to convince users in the West that their app is secure and in compliance with regulations that apply to them.

Another reason is that Super Apps have the issue of function creep. This happens when information is used for a purpose outside the original list of purposes agreed upon. It is often considered the greatest asset of Super Apps as this data can be used for a myriad of purposes. However, this reduces the trust in the App significantly and can be liable for compliance issues with data protection laws like GDPR. This then becomes the barrier of entry for Super Apps, unless they are able to set the guideline for what the data can be used for.

In the US, privacy has its roots in the very founding of the nation itself. With the controversies and laws in the US, people are more and more reluctant to take part in all-encompassing data collection. This concept of privacy and that only the parties directly and voluntarily involved should have access to related information heavily leans into the individual app culture in the West.

Case Study: Revolut

Revolut, headquartered in London, UK was founded in 2015 with its main focus on providing digital banking services in one app, with Nikolay Storonsky and Vlad Yatsenko being the founders. It started with 100,000 users in 2016 and now has 15 million users in 2021. It is known for its affordable services such as the lack of fees when exchanging money within a threshold based on the subscription used. It has four types of packages which are Standard, Plus, Premium, and Metal. It was originally launched in Europe but has steadily expanded into other markets since. It became a unicorn in 2018 when it was valued at £1.7 billion.

Some of its services include commission-free stock trading in NASDAQ and New York Stock Exchange, currency exchange, interest-bearing savings accounts or better known as 'vaults', debit cards, virtual cards, and more. Its subscription models do differ from country to country but most mainly have the Standard, Premium, and Metal. The benefits of each service differ for each package as well. Their standard package is free which includes basic benefits such as instant payment transactions, money exchange with the interbank exchange rate, spending abroad in over 150 countries, and setting up recurring payments.

During the pandemic, cryptocurrencies gained immense traction as an alternative to fiat currencies. The distrust in the central banking systems piqued the demand for crypto trading and pushed FinTechs such as Revolut to introduce its cryptocurrencies feature allowing users to buy cryptocurrencies with more than thirty currencies. Revolut is licensed as a Major Payment Institution in Singapore.

In April 2022, under the Payment Services Act (PSA) the Monetary Authority of Singapore (MAS) granted 'in-principle approval' to Revolut Technologies Singapore Pte Ltd (RTS) and Luno Singapore to offer digital payment token services. Revolut would offer fully regulated cryptocurrency services such as digital payment tokens, cryptocurrency, and others.

Revolut's Vault helps to save up for big events such as buying a home, saving up for further studies or marriage. Vault allows users to save for all these goals by simply rounding up their spare change, and transferring funds at any time such as one-time transfers or recurring transfers. Users can customize their vaults by choosing the saving amounts as well as the deadline that they wish to save up. To bring in the social aspect of the app, users can also create group Vaults with their friends or family to save up together in one Vault. The Vault also supports various currencies including cryptocurrencies as well as other commodities such as gold and silver. Revolut has also introduced the Saving Vault which has an annual interest of up to 0.65 per cent. This Saving Vault is only accessible to those who have a paid subscription. They are also able to withdraw from the Vault instantly without commitment or fees. The security feature of the Savings Vault is that it is protected by the Financial Services Compensation Scheme (FSCS) and deposited into their trusted partner bank.

Case Study: Uber

Uber is popularly known to be a ride-hailing app and now has the ambition to become a Super App. They have been taking baby steps to achieve their ambition such as the acquisition of several companies and diversifying their services.

Uber Eats is a food delivery platform launched by Uber in 2014 with a reach in more than thirty countries. Ever since its launch, Uber's strategic focus has been Uber Eats.

Even though Uber Eats made up 32 per cent of Uber's gross bookings in the fourth quarter of 2019, it accounted for 51 per cent of the gross bookings' growth. This trend did not just stop in 2019, continuing it into 2020 with the app contributing $1.6 billion in bookings compared to the previous year. So much so that Uber Eats is now bigger than its core service, ride-hailing.

In February 2021, Uber acquired Drizly, an online delivery store for local liquor stores. Uber has also acquired Postmates, an American food delivery start-up. According to Edison Trends in 2020, Postmates and Uber Eats have a 37 per cent share of food delivery sales in the United States. There are a total of fifteen acquisitions by Uber as of 2021. A number of acquisitions are delivery-focused such as Drizly, Postmates and Mighty AI, Swipe Labs, Geometric Intelligence, Otto, deCarta, Transplace, JUMP Bikes, Post Intelligence and RouteMatch which have their headquarters in the US, Cornershop with its headquarters in Chile, Autocab with its headquarters in the UK and Careem with its headquarters in Dubai.

Uber acquired Cornershop in 2019 which is an on-demand online grocery delivery application that operates mainly in Latin America. It delivers fresh produce, pharmaceuticals, and even pet supplies. from more than 3,000 stores. Uber wants to bank on a market that is projected to grow in value. This acquisition puts Uber as a strong contender in Brazil and other Latin American countries.

Although Uber has been heavily focusing on its food delivery platform, it has not forgotten its roots. In 2020, Uber acquired Autocab which is a British minicab software company. This acquisition allows Uber to offer services in places that it previously did not operate into. This includes Oxford, Doncaster, and Aberdeen. Uber saw that there were a large number of users that open the app in the cities where it wasn't operating about 25,700 in Exeter, 67,000 in Oxford, and 23,700 in Doncaster. Uber's partnership with Autocab is a strategic one, especially with the UK's heavily regulated market. Instead of going through the long process of legally setting itself up as a minicab firm in each town and city it operates in, it uses Autocab's partnerships with independent minicab operators. This streamlines the process of expanding its services to other areas. Uber has also acquired Careem to expand its ride-hailing services in the Middle East in 2019. In this deal, Uber acquired Careem's mobility, delivery and payment services across the Middle East region and major markets which includes Saudi Arabia, the United Arab Emirates, and Pakistan. Careem's market reach in the Middle East is massive with operations in more than 100 cities in fifteen countries and a user base of 33 million.

In 2019, Uber acquired Mighty AI, which develops training data for computer vision models. The purpose of this acquisition was to push for the development of self-driving cars. It will acquire Mighty AI's tech talent, tooling, and labeling community.

It is hard to see if there would be demand for Super Apps in the West like how it is in the Asian and Latin American markets. However, there would more likely be increased aggregation of services and an increased level of interaction between services. Big players that are typically the high-frequency used apps are now consolidating their services. This is evident in Facebook's ambition of becoming a Super App by increasing the interaction between the existing apps, such as Instagram and WhatsApp. The increased level of interaction between services can be done through partnership. This is where all partners can collaborate in a digital ecosystem to complete services efficiently and conveniently. For example, when searching for a movie ticket on Google, it shows all the options available and then pre-fills the form and brings the user to the check-out page within three clicks.

It is important to take note that the reasons we have discussed for the West's hesitancy apply to the East as well. Similar problems have been faced in the East in regards to Super Apps. The main reason why Super Apps have managed to come out as a leader in the East is due to the characteristics of its users in the region such as mobile first experience, less cultural barriers as well as limited concerns around data privacy and security. From the characteristics we have discussed previously, the way users perceive Super Apps is different in the East and the West. Each of those characteristics plays a part in making Super Apps a success in the East and not the West.

Even though it is hard to see the same level of demand for Super Apps in the West, with the continuous inventions being made, there is a chance for Super Apps to be in demand with better data protection. We have now looked in detail into the situation in the West in regards to Super Apps. Let's find out more about Super Apps and how it has flourished in the East!

Chapter 4

One App, One Tap

Did you know that the ASEAN Super App market is estimated at $4 billion in revenue and will have a projected increase to $23 billion in 2025? Well, in most Super Apps, the four usual services included would be ridesharing, food delivery, online banking, and e-commerce through FinTech.

Let's look at the structure of the Super Apps, its value prepositions, and dig in deeper to have a greater understanding and probe the question: Why are Super Apps a hit in Asia?

Some examples of Asian dominance in the Super App space include China's WeChat and AliPay, India's Paytm, Singapore's Grab, Indonesia's GoTo, Vietnam's Zalo, and South Korea's Kakao. Further in this chapter, get some answers on how Super Apps have been successful in the East and not so much in the West.

Structure of Super Apps and its Value Proposition
Super apps are the phone's ultimate go-to app
Because these apps are handy and save time, end consumers value the convenience of use and search of products and services in an all-in-one app. Consider it a clone of your home screen where you can access all of the services you need to organize your everyday life in one app!

Has a high open-rate for at least one service or function

Gojek, which is based in Indonesia, and Grab, for example based in Singapore, began as a ride-sharing app and then added features such as instant messaging and an e-wallet. GCash in the Philippines began as a mobile wallet for payments, branchless banking service, and a payment centre but has now expanded to include multiple verticals within the app. Whatever purpose a Super App was created for, it excelled at it, allowing it to evolve into an ecosystem of services.

End customers' wallets are easily accessible

Why should your customers keep their money in a different app when you have direct access to their wallets? Many apps with aspirations to be Super Apps provide this vital function, e-commerce platforms such as Lazada and Shoppe introducing their own e-wallet feature to make payments easier for their customers.

Partnerships with other platforms are encouraged and welcomed

Super Apps are similar to shopping malls, how are they similar to shopping malls? Well, they feature a variety of stores offering a variety of services. Their ability to be open to collaborations and partnerships is the exact reason they are who they are. Other platforms can be smoothly incorporated into the ecosystem they've created thanks to their app framework. Goama, a gamification platform, has relationships in over twenty-four countries and offers a carefully curated library of addictive games.

Super App's value proposition is to cover every online and offline demand of an internet user by replacing Amazon, Instagram, TripAdvisor, Booking.com, Venmo, Tinder, and PayPal with a single app.

Why are Super Apps Popular in the Asia-Pacific Region?

There are four key factors that make Super Apps more popular in the Asia Pacific as compared to the West, let's take a look at it, shall we?

1. The first factor is that there is high mobile and smartphone penetration. Countries such as Indonesia, China, and India have a high smartphone device ownership among internet users and a growing share of web traffic moving to mobile devices in recent years, with the percentage of smartphone users in the population approximately being 74 per cent in Indonesia, 57 per cent in China and 54 per cent in India. These three countries are among the top four largest smartphone markets worldwide, along with the US. The familiarity with mobile phones and the reliance on them allowed more services to be delivered through the apps.

2. The second factor is the significant unbanked population with limited access to basic banking services. According to the World Bank's latest report as of 2021, nearly 1.7 billion people are unbanked worldwide. That's close to one-fourth of the global population!

    Everyone in two people, over 70 per cent of Southeast Asia's population are unbanked or underbanked.[ix]

    But what do you mean by being underbanked? Well, being underbanked means that this sect of the adult population does not use banks or banking institutions in any capacity. Underbanked and unbanked are unable to access basic services such as cash withdrawal. This is why Super Apps in the Asia-Pacific region tend to target the less creditworthy, unbanked, or underserved people as they can brand the financial services as the most convenient way to do their services. Super Apps help to bring financial inclusion with its wide range of financial services. These include digital wallets (no minimum balance requirement unlike traditional bank accounts), insurance, payment of bills, small-size loans (offered without any credit score or history), and other services that can be done within the app itself.

3. The third factor is that numerous SMEs want to accept digital payments or are forced to do it due to their wide usage of it. It's apparent when you see stores that display a sticker that says what payment methods are accepted. Usually, they would display the typical payment methods like the brand of credit cards, but more and more stores are displaying Super Apps' logos like Alipay. If we focus on China alone, we can simply say that customers must either pay with Alipay or not pay at all as Alipay is the primary payment method everyone uses. There are 900 million domestic Alipay users alone and they spend 90 per cent of their day online on Alipay as well. Super Apps entered into a market where card providers had yet to develop a strong position. Cash was the dominating payment method in China before apps, allowing mobile to vault into first place. Alipay was also able to do something new and substantial because card schemes were not paying attention to the e-commerce space a decade ago. Businesses can accept payments by simply setting up a business account with Alipay. Coupled with government initiatives such as demonetization in India, acted as a precursor to the adoption of Paytm and BharatPe wallets.

4. The fourth factor is that there is less concern and awareness about data privacy issues. Consumers don't really get paranoid about data security and privacy. They are more focused on the benefits and convenience such as discounts or cashback. Valuation of personal data is low, with many people willing to pay only 1 per cent extra or nothing at all. People are not robust in their personal data valuation.[x]

Mergers and strategic partnerships help to expand offerings beyond the core capabilities hence capturing a bigger pie of the market share. Let's look at some instances, shall we? GoTo (earlier GoJek) launched GoPlay, a video streaming service by joining hands with local and international studios. Ant Financial, in association with Vanguard, offers investment advisory services to its retail consumers. Users are hooked and locked to the pandora box as Super App keeps throwing surprises and new features.

Let's look at the state of specific Super Apps that are performing excellently and are on their way to make it to the top, or have already reached that milestone. We shall dig deeper and look at their timeline, their strategic offerings and how are they holding up in the market right now.

To get a more in-depth view and understanding of the current scenario of some Indian Super Apps, I sat down for a chat with the founder of PhonePe, Sameer Nigam.

State of India's Super App Scenario

PhonePe

But firstly, is PhonePe even a Super App? Or is it going to be one soon? According to Sameer Nigam—CEO of PhonePe—

the main goal of PhonePe is nothing close to what a Super App essentially is, Sameer's way of describing it is, 'An app with a super wide array of services being offered'. PhonePe's mission at its core is to provide a digital payment platform, if there is a digital market, there is a digital payment platform as well. PhonePe has captured a large audience 'We've captured half of the smartphone market in India,' shares Sameer when asked about inclusion.

But why isn't PhonePe a Super App? It's a pretty big platform and has the potential of being one. PhonePe is doing extremely well with providing basic digital payment solutions and during the interview, Sameer also mentions, 'we have pretty much every bill in the platform'. What PhonePe lacks are banking features, PhonePe pretty much sticks to its digital payment idea and is doing exceptionally well in achieving their goals which we will see in detail further on in this chapter. Pertinent to mention that Paytm made a foray into the banking space through its Payments Bank in 2017, a mobile-first bank with no minimum balance requirement and zero charges on all online transactions. As of March 2022, the Reserve Bank of India (RBI) directed the bank not to onboard new customers for allegedly violating India's data storage rules. This is a recent backlash in the spree of issues the company is already grappling with.

A way of attracting more consumers to a new business is obviously giving them excellent service but some companies go down the route of 'cashback'. Sameer however does say that they offer cashback in the first transaction 'but if I have to carry on doing that each time, you're not using this service, you're trying to make money doing things that you ordinarily may not have used.' The goal here is to not only bring in business but it's to really sow the seed of 'digital payments' in the minds of consumers leading them to a whole new ecosystem which is more convenient for not only people in urban areas but also people in rural areas where a lot of services are not very readily available.

PHONEPE HISTORY
THE TIMELINE
2015–Present

2015
December
FOUNDED BY
SAMEER NIGAM,
RAHUL CHARI,
BURZIN ENGINEER

2016
Acquisition by Flipkart
AT VALUATION
BETWEEN
$10MILLION–$20MILLION

2017
First payment app
reach 10 million downloads in
demonetisation

2016
First payment app
using UPI during
demonetisation

2018
Partnered with Wallet Players :
Freecharge, JioMoney, Airtel
Money-showing users to link
with existing platform Phonepe

2019
PhonePe records phenominally
exponential growth rate of
857.2%, surpassing GooglePay
market share.

2020
January - June
PHONEPE INTRODUCES NEW
FEATURE TO ALLOW USERS
TO WITHDRAW CASH
WITHOUT USING CARD

2020
January
PHONEPE IN
COLLABORATION WITH
BAJAJ ALLIANZE INSURANCE

2021
PhonePe recorded over
125 million
active users per month

2021
PhonePe has 1.3billion local
transactions per month (over
1 billion UPI transactions per
month on platform.

PhonePe is one such payment app that has made the lives of millions of Indians easier. Users in India can link their credit and debit cards to a mobile wallet and make digital payments.

Founded in December 2015 PhonePe is the first payment app built on the Unified Payments Interface (UPI). But what is UPI? The textbook definition of UPI states that Unified Payments Interface (UPI) is an instant real-time payment system developed by the National Payments Corporation of India (NPCI) an interoperable infrastructure for inter-bank peer-to-peer (P2P) and person-to-merchant (P2M) transactions. In 2016, PhonePe was acquired by Flipkart.

Things quickly changed with PhonePe's entry on UPI platform. In the early days of UPI in 2016, Paytm, MobiKwik, and Freecharge were leveraging their payment infrastructure. Demonetization disrupted the digital payments industry in India. UPI transactions went sky high. When PhonePe was launched in 2016, the idea was to facilitate all types of transactions between customers and merchants rather than just payments, PhonePe currently has around 22 million merchants associated with them. PhonePe became the first UPI-based app to reach 10 million downloads in 2017.

PhonePe enables its 20 million+ of merchants (offline + online + in-app) to accept UPI-based digital payments through interoperable QR codes. Merchants accept PhonePe across 12,000 towns (of 17,000 towns overall in India) and 4,000 talukas or groups of villages (of 5,500 talukas overall in India). The merchant partners are offered a personalized store page on the PhonePe app wherein they can list store timings and share their product catalog. It also enables merchants to offer home delivery by facilitating remote payment from the app.

Fast forward to 2019, PhonePe dominated the year, growing at a rate of 857.22 per cent, 335 million registered users, with over 145 million active users. PhonePe surpassed its competitor, Google Pay, in terms of market share. In May 2021, PhonePe had 125 million active monthly users and 2 billion total monthly transactions (Over 1 billion UPI transactions) on the platform.

PhonePe became the fastest-growing InsurTech player in India. Insurance companies adopting tech is called InsurTech—easing the process of onboarding and settling the claims. PhonPe disrupted the decade-old insurance space by offering COVID-19 insurances by the name of 'corona care' which is an effective alternative to health insurance. PhonePe is also leading the digital gold business with a 35 per cent share. Mutual fund, which was not that big a category for the company till the beginning of 2020, has grown to over ₹100 crores in Assets Under Management (AUM).

PhonePe introduced a new feature in 2020, which allowed users to withdraw cash from their bank accounts without using a card. The PhonePe ATM is a digital ATM that allows users to withdraw cash by partnering with merchants. The service is completely free, and the user will not be charged for using it. The withdrawal limit will be determined by the user's bank. A customer in need of cash can simply open the PhonePe app, navigate to the Store tab, and click on the PhonePe ATM icon to find nearby stores that provide this service. Customers simply need to go to the nearby shop, click the withdraw button, and use the PhonePe app to send the appropriate amount to the merchant. The merchant will give the consumer cash, equivalent to the amount transferred. The PhonePe ATM is available in 10 lakh locations across 300 cities in India. With more than 2,200 employees PhonePe is roaring like a tiger and supporting the mission of Digital India to bridge the financial and digital divide.

PhonePe Business Model

Customer Value Proposition

"Never Run Out of Balance"
First UPI-based Transaction Application in India
Acquired by Flipkart. India's leading Ecommerce Giant in 2016
Leveraging on Easy Availability of Internet and Smart-phones
Now, a Daily Utility Free App in India

Online Transactions

Facilitating India through Fund transfers during Demonetisation, P2P, Self Bank Transfers, PhonePe Wallet

Utilities

Recharge Mobile, FasTag and Pay Bills to Electricity, Water, Gas, Credit Card, Rent, Postpaid, DTH, Broadband

Travel Metro Recharge/ QR

Recharge Online on Metro smart card or Book QR Tickets Online, tap on AVM, and go.
Delhi, Mumbai, Hyderabad, Bangalore

Financial Services & Taxes

Credit Card Bills, Loans Repayment, LIC/Insurance, Municipal Taxes

Revenue Model

The Business Strategy: Commission-Based Payment Gateway: No.of Transactions made

Insurance

General Health, Health+, COVID-19, COVID-19, Motor, Travel, Accident, Term, Guaranteed Returns, Life : Facilitating through Online Transactions, Earning Commissions

Switch Apps

"A shortcut to a world of Apps"
Orders/Booking - Trscking & Modifications, Cancellations & Refunds, Offers & Cashback

Investment Ideas - Digital Gold

24k Gold Coins from 0.5g to 20g, Silver Coins from 10g to 100g

Lenskart, IRCTC Trains Tickets & eCatering Food on Track, Swiggy, RedBus, Ola, Uber, Oyo, PharmaEasy

PhonePe Research Team

PhonePe Research Team suggests Tax Saving Funds, Mutual Funds, SIP & Growth Funds, Saving Funds, Duration Funds, Csp & Debt Funds

Paytm

PAYTM HISTORY
THE TIMELINE
2010-Present

2010 August
FOUNDED BY VIJAY SHEKHAR SHARMA AS PREPAID MOBILE AND DTH RECHARGE PLATFORM

2013 January-June
ADDED DATA CARD, POSTPAID MOBILE AND LANDLINE BILL PAYMENTS

2014 January
PAYTM WALLET LAUNCHED AND USED BY INDIAN RAILWAYS AND UBER PAYMENTS DONE BY MOBILES

2013 October
INVESTMENT OF $10 MILLION BY SAPPHIRE VENTURES

2015
Provided payment services from education fees, metro recharges, electricity, gas to water bill payments due to demonitisation

2017
India's first payment app to surpass 100 million downloads Paytm launches Paytm Gold, Paytm Bank,Paytm mall, Paytm business

2019
Paytm launches Paytm First Credit Card, in collaboration with State Bank of India

2018
Paytm launches Paytm Money —SEBI registered Investment Adviser (IA)

2020
Paytm introduced the option to buy car and bike insurance.

2021
Paytm added equity trading allowing users to buy and sell stocks through Paytm associated trading accounts

Paytm is an acronym for 'Pay through Mobile', Paytm was initially launched in 2010 as a website for a prepaid online mobile recharge facility. Slowly they started adding various features like online bill payments. In January 2014, Paytm Wallet was

introduced on Indian Railways and Uber as a payment option. It
soon took a plunge into e-commerce, bus ticketing, metro top-
ups, utility bill payments, etc. The turning point for Paytm was
the demonetization, when ₹500 and ₹1,000 were discontinued,
the general public was forced to turn towards e-wallets, and this
helped Paytm amass more customers.

In 2017, Paytm launched a consumer shopping app for its
users, known as Paytm mall. The model drew inspiration from
China's largest business-to-consumer (B2C) retail platform. Their
aim was to create a combination of the mall and bazaar concept
for Indian consumers.

With the help of the new app, consumers were able to shop
from 1.4 lakh sellers. Paytm also started offering users the facility
to buy 24K gold through their app. The buyer can store the gold
in a secure locker, deliver it through a safe channel, or directly sell
it through the Paytm app.

At the end of 2017, Paytm along with its parent company
launched a Payment Bank. The payment bank allowed the user
to have and utilize a savings and current account with Paytm,
wherein they could withdraw and deposit money into the account
and use Paytm-issued debit cards linked to their accounts.

Paytm launched Paytm money in 2018, Paytm Money Limited
is a registered Investment Adviser (IA) that offers investment
execution & advisory services as per the Securities and Exchange
Board of India (SEBI)—regulator for securities and commodities
in India. The feature allowed users to invest in mutual funds as well
as national pension schemes. In 2021, they added equity trading
allowing users to buy and sell stocks through Paytm associated
trading accounts. Paytm introduced the option to buy car and
bike insurance in 2020. They also started to offer credit cards in
collaboration with the State Bank of India.

Paytm tweaked its business model wherein the customer
buying the subscription (Paytm First) will receive exclusive

benefits as well as cashback. Paytm, has been fighting tooth and nail with PhonePe and Google Pay to retain its leadership position in its core areas such as mobile payments and financial services.

Other products like Paytm Mall and its payments bank that were supposed to take off and could have blended well into its Super App strategy did not perform as expected. Another reason the service hasn't taken off is that it launched too many of them too quickly, without giving its 400 million-plus registered users enough time to understand and become acquainted with the new offerings. Paytm also once received unicorn status which it very recently lost, the FinTech company has been incurring losses for years- huge carry forward losses and negative earnings.

Though Paytm was the first app to become a Super App in India, it has been beaten by emerging competitors in the payment space as well as forming strategic partnerships with service industry leaders. Paytm needs to offer something different than its competitors in order to retain its title of India's successful Super App.

Tata Super App

Tata Group was founded in 1868 by Jamshedji Tata, India's biggest and oldest industrial group, catering to FMCG and the service sector. The company gained international recognition after purchasing several global companies.

Despite being active in a variety of industries and owning major retail brands across India, they had a very small online presence in comparison to their competitors. Tata was facing the same problem as Walmart, they realized that they did not have personalized data to enhance customer experience compared to Amazon and Flipkart.

To address the problem, in 2019 Natarajan Chandrasekaran, Chairman of the Board of Tata Sons, reorganized the thirty listed companies and 1,000 subsidiaries within Tata Group. The companies were divided into the following ten categories: IT, Steel,

Automotive, Consumer & Retail, Infrastructure, FS, Aerospace & Defence, Tourism & Travel, Telecom & Media, Trading and Investments. The cornerstone of the regrouping was Tata Digital app, with the aim of making products that will enrich the lives of customers and also leverage their mass distribution network.

Tata's decentralized ways of working have been both a boon and a curse throughout its history. It enabled the establishment of new businesses with their leaders, but it also resulted in a loss of control. Brick-and-mortar brands such as Westside, Titan, Tanishq, and Croma had both an offline and a limited online presence via their own websites and e-commerce platforms. Bringing them together could aid Tata in the development of an e-commerce platform.

The group's first attempt into e-commerce was the creation of Tata CLiQ, with Q symbolizing the brand's focus on only the best brands and products for consumers with impeccable taste. To set itself apart from its competitors, it adopted a no-cashback policy from the start. However, because of its primary focus on the luxury segment, Tata CLiQ lagged behind in the race to compete with the likes of Flipkart and Amazon, which had captured the market due to their appealing high discounts and cashback on products sold.

Despite growing e-commerce penetration, Tata CLiQ's investment by the end of 2019 was still a fraction of what Amazon and Walmart had. Tata CLiQ transitioned from a marketplace to a direct seller with its own inventory in 2020. The inventory model was also implemented in the hopes of increasing margins through lower logistics costs. This was possible because, unlike their foreign competitors, Tata was not bound by complicated e-commerce regulations for FDI. Tata CLiQ received a fund infusion of ₹3,500 crores in January 2021, which was nearly twenty times what it received in 2019. Setting ambitious growth goals for the future, and setting Tata CLiQ up to be a Super App.

According to experts, Tata will look to leverage its nationwide offline presence to boost e-commerce. This would entail exchanging customer data from new businesses and comprehending concepts such as customer overlap and cross-selling.

Tata Digital has been tasked with developing this platform over the last few months. Under the new Super App platform, the Tata group intends to offer a wide range of products and services including e-commerce, financial services, fashion, and lifestyle, among others.

It is rumoured that Tata Digital would acquire a significant stake in the hyperlocal delivery platform Dunzo. Bringing in Dunzo could help Tata expand their local footprint. Financial services such as loans, insurance, and mutual funds are also anticipated to be offered by Tata Digital. To provide financial intermediation, Tata Digital is looking for strategic partnerships with regulated banks and insurance firms. Credit card applications, insurance distribution, microloans, and even merchant management are examples of these services. Tata Digital might also buy a neo-bank or try to set one up themselves.

Tata Digital is eyeing global investors to raise $2.5 billion in order to launch the Super App. Tata Sons Chairman N. Chandrasekaran had said in an interview in late 2019 that the Tata Digital platform would be launched by May 2020. While the pandemic did push that deadline but now it's confirmed that the Super App, Neu will be out on 8 April 2022. Since Tata is the title sponsor for the Indian Premier League (IPL) cricket, the announcement will come around then. The app promises to bring a unique reward system for its users.

There is no denying that no other company has the same impact on Indian consumers as Tata group. Its legendary culture, relentless innovation, and proclivity for autonomy will be its pillars to transform the digital landscape within India. In the race to become India's first successful Super App, Tata has all the ingredients to be your one-stop-shop for everything.

Reliance Jio

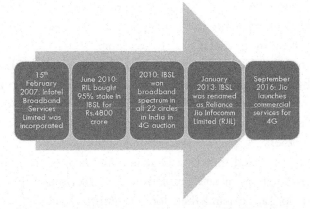

15ᵗʰ February 2007: Infotel Broadband Services Limited was incorporated

June 2010: RIL bought 95% stake in IBSL for Rs.4800 crore

2010: IBSL won broadband spectrum in all 22 circles in India in 4G auction

January 2013: IBSL was renamed as Reliance Jio Infocomm Limited (RJIL)

September 2016: Jio launches commercial services for 4G

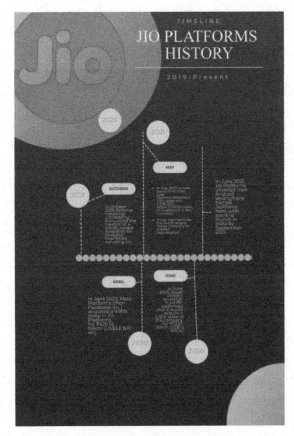

Reliance Jio is a soon-to-be Super App. It began a concerted effort to acquire start-ups in its areas of interest in order to strengthen its capabilities and realize its vision of 'Digital India'. It established the Jio Digital India Start-up Fund in September 2016 for ₹5K cr ($750M). It is stated that Reliance Jio's goal was to create a platform for young Indians to create 'future businesses.' Reliance Jio encouraged the young generation to develop applications that were compatible with the Jio app, a strategy inspired by Google's decision to open up its Android OS to allow developers to innovate on top of it. Though initially, customers had to download each app individually, it can be considered a minuscule step towards being a Super App.

Reliance Jio did what every tech giant will do. Within a span of two years, it acquired or invested in companies or start-ups from various sectors but it's been unable to monetize any of these platforms. Its big hits, such as Saavn (an Indian music streaming service and regional digital distributor) and Embibe (educational platform providing free learning content powered with artificial Intelligence personalized recommendations), did not provide benefits as expected. Reliance appeared to have overinvested in the diversification game. Its in-house brands were in the same boat. While both Reliance Digital (gadget store) and Fresh (grocery store) were increasing their revenue share, the benefits of becoming a traditional market leader were limited. They were marketed as Reliance's first technology offerings, but they were buggy and provided a poor customer experience.

Then Reliance Retail went on a shopping spree in order to dominate the market. It established a presence in over 6,600 cities and towns, opening stores at a rate of ten per day. Subsidiaries such as Reliance Retail were offering massive discounts in order to compete with behemoths such as Amazon and Flipkart.

Reliance Jio had successfully established two parallel streams, a) offline muscle and b) online presence. Reliance had significant offline clout and online capabilities throughout its long history, but none of that would help it become an online behemoth. The time had come for a non-traditional tech innovation that could combine its catalogue of offline and online consumer products into one.

Then came the opportunity to form a partnership with Facebook, which would grant access to 400 million WhatsApp users. Sellers and buyers would communicate via WhatsApp. Participants will be able to place orders on a platform with which they are already familiar. WhatsApp would fill a critical gap in Reliance's technological pivot. JioMart, Reliance's own e-commerce platform, had already been launched. Within days of Facebook investing in Jio, JioMart was officially launched on WhatsApp.

But there is something very different about Reliance that brings to light another aspect of Super Apps, which is unlike other e-commerce behemoths: this new digital platform will not replace physical stores (Kirana stores). Instead, it will work with them to expand both their reach and profitability. This is a very smart way of integrating 30 million small and medium-sized businesses (SMBs), this could be extremely beneficial to customers as well as businesses.

Reliance Retail Ventures has acquired the furniture e-tailer Urban Ladder as well as Vitalic Health Private Ltd, which specializes in pharma distribution, sales, and business support services, as well as the online pharmacy platform Netmeds. JioMart that began as a grocery delivery platform, is now expanding into other categories. Reliance has a retail network of over 12,000 stores. They recently invested in Justdial, a B2B and B2C platform in India. It also owns the iconic British toy retailer Hamleys and has exclusive rights to sell a number of luxury fashion brands in India.

Reliance has not publicly stated its intention to create a Super App. However, Reliance's MyJio app has already evolved into a one-stop-shop, allowing the user to manage Jio mobile and fibre services, pay bills, transfer money via UPI, stream movies, music, and games, and even connect to JioMart.

The introduction of the mini-app is another step towards being a Super App. In 2020, Google invested $4.5 billion in Jio Platforms taking a 7.73 per cent stake in the company. This was Google's first investment through its India Digitization Fund ($10 billion to accelerate India's digital economy).

In July 2020, Google announced another $337-million investment to develop a low-cost 5G device, a Jio smartphone, and a new 5G collaboration powered by Google Cloud. In addition, the company announced the adoption of Google Cloud for several Reliance businesses. Jio will use Google Cloud to power Jio's 5G solutions, as well as Reliance Retail, JioMart, JioSaavn, and JioHealth. Reliance will migrate its core retail operations to Google Cloud infrastructure in order to scale up and improve customer experience. Google Cloud will also provide a complete end-to-end cloud offering for Jio's 5G network and services, allowing for fully automated lifecycle management. Furthermore, Jio and Google Cloud will collaborate to deliver a portfolio of 5G edge computing solutions to assist industries in addressing real-world business challenges.

Jio will look into developing new services in the gaming, healthcare, education, and video entertainment industries. This is an opportunity for Jio to promote their Super App and solve the shortcomings associated with affordable smartphones.

With Google and Facebook on board, as well as a new strategy to revitalize dormant apps, Jio Platform has the potential to reshape India's digital economy. Being able to launch a low-cost smartphone loaded with all of the utilities powered by Google Cloud and aided by a homegrown mobile network makes Reliance Jio a strong contender for launching a successful Super App in the future.

Monopolization of Super Apps in China

Super Apps have invaded every aspect of digital life in China. The unique feature of Super Apps is high daily usage rate. In China, people spend more than 90 per cent of their time on a single app such as WeChat or AliPay. The mobile first experience got them hooked to their favourite app as the one stop solution for messaging, social media sharing, ride hailing, payments et all. WeChat is the world's first Super App. If you are curious on how they got to where they are today, let's take a look:

WeChat

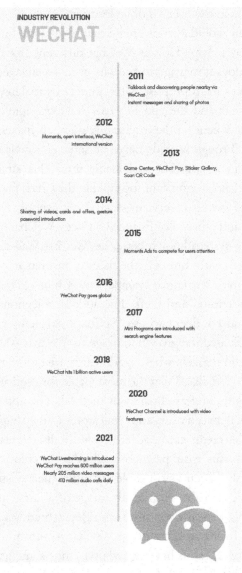

INDUSTRY REVOLUTION

WECHAT

2011
Talkback and discovering people nearby via WeChat
Instant messages and sharing of photos

2012
Moments, open interface, WeChat international version

2013
Game Center, WeChat Pay, Sticker Gallery, Scan QR Code

2014
Sharing of videos, cards and offers, gesture password introduction

2015
Moments Ads to compete for users attention

2016
WeChat Pay goes global

2017
Mini Programs are introduced with search engine features

2018
WeChat hits 1 billion active users

2020
WeChat Channel is introduced with video features

2021
WeChat Livestreaming is introduced
WeChat Pay reaches 800 million users
Nearly 205 million video messages
410 million audio calls daily

WeChat is one of the most successful Super Apps boasting of an estimate of over 900 million users. Over a third of the users spend more than four hours per day on the app. It started off as a messaging app but soon gained momentum to cater to different touchpoints in the user's mobile journey.

But why would anyone spend that much time on an app even if it's a Super App? How is WeChat different from other apps? And how does it manage to keep its users so engaged?

WeChat's main purpose is to connect by making connections between users and the app in many ways possible regardless of wherever or whenever the service is required. It makes connections with users through its wide range of services, slowly entering into users' life in a slow but sustainable way. This strategy allowed WeChat to gain a significant foothold in the China market. WeChat consists of various services such as financial services, retail platforms, and others. For financial services, WeChat has developed its payment system known better as 'WeChat Wallet'. The system is partnered with one of the largest insurance companies in China, People's Insurance Company of China (PICC), to protect users against fraud and theft. The added protection makes users feel safe and confident to use WeChat's payment services which further facilitated the growth in usage of WeChat Wallet. It targets its services in markets where credit cards are not commonly used. This makes WeChat Wallet the most viable for users in that market to use it as its preferred payment method. The app allows easier connection between users and retail stores as they target users who don't hold a credit card and are typically disadvantaged as most online platforms' main payment method is the credit card.

Not only are retail stores able to capture their existing audience, but they are also able to take advantage of a large audience from Super Apps to sell their products. Retail brands can create an official and verified account on WeChat to set up a shop within the app. If the brands have signed up to the payments Application Programming Interface (API)—a software intermediary that lets two applications talk to each other. This way users need not leave the app to complete their purchase. The API is known as Quick Pay which has a range of services for merchants to use. There are multiple APIs under Quick Pay such as Submit Quick Pay, Query Order, and Query Refund. The range of services includes details of settled funds, exchange rate queries, payment orders made from WeChat as well as option to download reconciliation files.

Not only does this bring convenience to users, but it also achieves Super App's goal of connecting and integrating into users' lives. There are other features of WeChat that make it the behemoth of a Super App today. This includes stickers, QR codes, and iBeacons (integrate its services with offline platforms). Stickers are one of the favourite functions within the China market with their wide range of choices for various purposes. Stickers are something that seems trivial but are a huge part of users' daily life where they commonly use them to express their emotions when chatting with their friends and family. WeChat originally started out as a messaging app, so it is only natural that one of their core services' features would be widely liked by users.

Another feature of the messaging services would be its QR code functionality. This allows users to create their QR codes to connect with others without saving their contact. This is extremely convenient for users as they can communicate directly without extra steps. WeChat creates technologies like iBeacon to integrate its services with offline platforms. Better known as omnichannel retailing, which refers to the integration of online and offline retail platforms. This strategy is often used to increase sales and enhance customers' in-store experience. An example of this would be that WeChat allows users to pay for bills or products on their phones and collect their products by showing their QR codes to the staff in-store for verification. These technologies can also help to reduce operational costs. Fewer resources are used as the app performs most of the services. A significant example of this would be the Chat Inn, a WeChat Smart Hotel. Users can complete most of their services using the app such as booking a room and ordering meals. It is also efficient as processes such as checking out of the hotel can be done almost instantaneously with a single tap. WeChat offers instant translation of messages, dating, and social media similar to Facebook. This action of a single tap is what most users are looking for when using services and what makes Asians dependent on the Super Apps.

AliPay

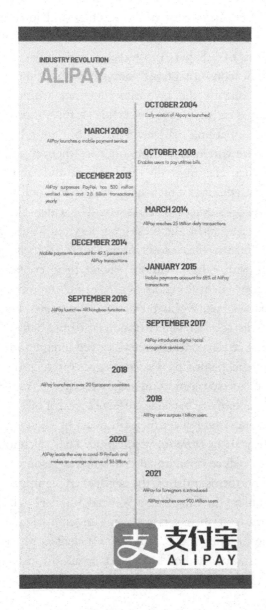

Alipay was developed by Alibaba in 2004. It leads the global mobile payment platform market although since then Alipay has faced stiff competition from WeChat and others. Alipay and WeChat jointly command 90 per cent of mobile payments in China. In 2019, it had an average of more than 320 million daily active users.

Street vendors, mom-and-pop stores love Alipay, with the QR code scanner feature they don't need hardware installation and anyone can make a purchase in the blink of an eye. In the offline stores, merchants can choose to accept Alipay payments via POS terminal, QR code, cash register software, and the mobile app.

Alipay accelerated in the Chinese market due to the low penetration rate of credit cards. Hence, in place of credit cards, mobile payment has gained significant market share instead. The mobile payment offers a frictionless, easy, and fast user experience which is something the users appreciate. It has also become the preferred payment method for Chinese tourists. Insights from Nielson's survey show that mobile payment usage by Chinese tourists has increased to 69 per cent in 2018 while the use of offline payment methods such as cash and bank decreased to 7 per cent. Not only is it a convenient method for Chinese tourists to pay for goods, but it is also beneficial for overseas merchants where they have experienced increased store traffic. In that same survey, more than half of the merchants in Singapore, Malaysia, and Thailand said that their store traffic has increased after implementing Alipay. Such surveys weren't carried out post-2019 as COVID-19 put the world on a halt, let alone travelling and spending by tourists.

Other Super Apps from Southeast Asia

Southeast Asia is proving to be a hotbed for Super App innovation. With the economy set to become world's fourth largest by 2030, with 57 per cent tech savvy population under the age of thirty-five, Southeast Asian apps are set to dominate the Super App ecosystem. Grab, is the cherry on the cake with its enviable journey to the top slot.

Grab

Grab was founded by Harvard Business School alumnus Anthony Tan and Hooi Ling Tan in 2012. Ling Tan was on a mission to make the commute safer for women in Malaysia. She recalls calling up her mom every time she would flag a cab, hoping sharing the taxi number and driver being privy to the conversation would make the journey safer. After securing a $25K grant from their business school, the duo started MyTeksi in Malaysia, which was later rebranded as 'GrabTaxi' in 2013. Soon, Grab Holdings was providing numerous consumer services, including ride-sharing, food, package delivery, and even online payments. It is one of the most valuable start-ups in Southeast Asia with a valuation of $16 billion dollars. As of now, it operates in eight countries in Southeast Asia namely, Malaysia, Singapore, Indonesia, Thailand, Vietnam, Philippines, Myanmar, and Cambodia. In terms of competition, Grab faces global competition from Uber, Didi Chuxing, and Lyft, whereas in the Southeast Asian market it stands against Gojek and Mai Linh (Vietnam).

In 2016, the company rebranded itself from a taxi-hailing service to providing deliveries, digital payments, and insurance. In 2018, Grab launched GrabFood and also invested in digital payment service OVO to compete with Indonesian rival GoTo. Grab Holdings now runs numerous subsidiaries such as Grab Transport (taxi, private car, motorcycle, and ride-sharing services), GrabFood (food delivery), GrabMart (essentials delivery), GrabInsure (insurance), GrabRewards (rewards programme), Grab Express (parcel delivery), GrabPay (online payments), and GrabGifts (gift cards).

The Grab ecosystem integrates all needs in a single go. The app allows users to pay for taxi bookings and settle payments through GrabPay. Likewise, it allows users to buy groceries using GrabMart and book a cab simultaneously. GrabPay acts as a digital wallet where users can perform cashless transactions by linking their credit/debit cards. Grab is said to have around 25 million monthly active users and has clocked in 1.9 billion transactions.

Grab Financial Group sold over 100 million insurance policies across Southeast Asia by February 2021. Grab Invest is a Robo investing platform that allows investment starting as low as $1. Likewise, Grab also plans to venture out to new areas such as fraud detection and targeted advertising for its drivers and retail partners.

GoTo

Go-Pay is Indonesia's fourth-biggest e-wallet service, the top three are namely, Bank Mandiri's e-Money, Bank Central Asia's Flazz, and telecom firm Telkomsel's TCash.

Kevin Aluwi and William Tanuwijaya, thirty-somethings are Indonesia's most successful founders. They are the founders of GoTo. The Indonesian tech giant GoTo Group is a merger between two of Indonesia's largest start-ups Gojek and Tokopedia. Tokopedia was set up right after the financial crisis in 2009, an e-commerce marketplace to connect small merchants with buyers while Gojek started in 2010 as a ride-hailing platform. Both businesses took the opportunity of the rapidly growing middle-class population and ventured into digital payments and other services in the following years. Tokopedia added parents and small stallholders to its e-commerce platform. Gojek regionally expanded its ride-hailing platform and turned it into a Super App by offering users on-demand services from food to massages and pedicures. In 2015, the two companies partnered up where Tokopedia would use Gojek drivers to provide same-day delivery for their products during off-peak hours. This was reportedly the first partnership between an on-demand platform, financial services and an e-commerce platform in the world.

In the midst of growing competition between tech companies regionally and globally, the pair officially merged in 2021 valuing at $18 billion dollars. Indonesia is the world's fourth-largest populated country, the joint entity represents 2 per cent of Indonesia's $1 trillion GDP, combining over 100 million monthly active users with more than 11 million merchants and 2 million drivers. Currently, GoTo offers three different products and services. Gojek is an on-demand multi-service platform that books transportation for drivers or for grocery deliveries. Tokopedia is an online shopping portal and GoTo Financial contains services such as GoPay, GoSure (insurance), GoInvestasti (investments), GoPayLater (a buy now pay later service), Midtrans (payment gateway), and Moka (point of sale).

Although a lot of their early growth was driven by the urban centres and Java, GoTo is planning to expand its horizons and reach out to the 120 million Indonesians that live outside these urban areas and also including the 17,000-island archipelago. This group of people includes 47 million adults who lack mainstream financial services and products, and 92 million people who have never used a bank. GoTo is looking to tap into this unbanked and underbanked sector and offer them basic financial services and create an impact.

Ahead of the company's IPO, GoTo raised $1.3 billion from Google, Tencent, Temasek (SG), Malaysian sovereign wealth fund Permodalan Nasional Berhad and a wholly-owned subsidiary of the Abu Dhabi Investment Authority.

Kakao

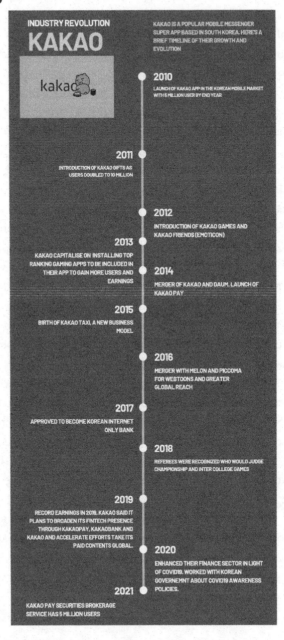

Kakao was established in 2010 in South Korea, offering KakaoTalk, KakaoTaxi, and KakaoMap. In 2021, Kakao became the third most valuable company in South Korea. Its current valuation as of June 2021 is $57.4 billion and its year-to-date stock price growth is 85 per cent.

One of its features, KakaoTalk was launched in 2010 as a messaging app that has more than 200 million users and more than 40 million monthly active users' equivalent to 88 per cent of South Korea's population. Some services offered by KakaoTalk include Kakao Shopping, Kakao Channel, Kakao Mail and Kakao Wallet.

Kakao merged with Kakao Commerce to compete with the industry's leading players like Naver in 2020. In the same year, Kakao Commerce's gross merchandise value has an estimated value of $447 million. There is also a planned merger with the fashion platform, ZigZag, which was acquired for $900 million in April 2021. Kakao is certainly looking to take advantage of the growing market of e-commerce in Korea. From messaging app to e-commerce, seems like the right thing to do for the Super Apps.

Kakao Commerce launched its live commerce service called Kakao Shopping Live in 2020. It takes inspiration from a traditional home shopping channel and turns it into mobile service. Sellers are provided with tools and techniques to produce content easily. To notify customers of the live broadcasts, Sellers are also able to communicate with the customers on the app as well to clarify any doubts that customers might have. Kakao Shopping Live's KakaoTalk channel has more than 1 million subscribers. 'Talk Deal' is a new addition to the platform that sells goods at discounted prices for group buying. Users can open their own deals and join others to take advantage of the discounts, even if they are strangers to each other. Users that open deals receive Kakao points which is equivalent to 2 per cent of the payment and the participants of the deal receive 1 per cent of the payment. KakaoTalk Gift service is a popular service that allows users to send gift coupons to others through the app. In 2017 alone, transaction volume exceeded 1 trillion won and is the lead player in the online gift market.

The app also has its own banking feature called KaKao Bank. Kakao Bank is a South Korean neo bank (no brick mortar) established in 2016 after it won the online banking license from the Korean government. The bank's initial public offering (IPO) in Seoul was close to 50 per cent larger than the established financial institutions such as KB Financial group. On its trading debut, its stock shares surged by 80 per cent with its market capitalization of more than 32 trillion won, making it the first internet bank or company to be the top lender in the country. In 2021, it accounted for more than 14 million customers. They are also able to take advantage of the big data collected from KakaoTalk which is then used to customize the product offerings based on the consumers' behaviour on the app. Consumer behaviour may include spending decisions, personal preferences, and design choices.

Its services include loan deposits, debit cards, and others that are performed entirely on the smartphone. There are a few types of deposit products which include instalment-savings accounts and time deposits. In an instalment-savings account, users have the flexibility to adjust the maturity and instalment amount. Its loan journey is entirely online without the need to go into a bank. Its loan services include secured and unsecured credit loans, prime credit loans, overdraft loans, and more. The bank also provides other services such as group account service, loan platform service, and credit information service.

As Kakao does not need to worry about operating offline platforms such as bank branches, it focuses on in-app experience. Hence with its heavy investment in user experience, Kakao Bank prides itself as an app that is easy to use. Users can apply for an account within seven minutes via a simplified application process. Users can manage their financial health via viewing their credit score information, balances of credit and debit cards, and loan status free of charge. It also uses biometrics for users to easily log in to their accounts. Kakao Bank's main two strategies would be its convenience and competitive pricing. The convenience comes from the user experience of the app. Competitive pricing refers to the high-interest rates for deposits and lower interest rates for loans as compared to its competitors. It also scraps fees such as withdrawal fees and early redemption charges to ensure that its services are affordable for others.

Kakao Bank has also partnered with various financial institutions for products and other purposes. Users now can open a stock trading account on its platform through its partnership with Korea Investment & Securities and NH Investment & Securities. Kakao Enterprise has also partnered with the Bank of Korea (BOK) in 2020 to utilize the BOK's AI technology capabilities as part of BOK's initiatives for digital transformation by 2030.

In June 2021, Kakao Bank shared that it is planning to increase the exposure to less creditworthy customers. This is done through adopting a new credit-scoring system that takes in various data concerning both financial and non-financial sectors to determine which users are more likely to pay back their loans, even if their credit score is low for other reasons. In March 2021, it also raised its upper limit of loans from $45,000 to now close to $60,000.

Kakao's story as a financial platform has just begun, with many more laurels to achieve after their stellar IPO. We will look deeper into the Initial Public Offer (IPO) stories of these Super Apps in the coming chapters.

Zalo

INDUSTRY REVOLUTION

ZALO

ZALO IS A POPULAR MOBILE MESSENGER SUPER APP BASED IN VIETNAM. HERE'S A BRIEF TIMELINE OF THEIR GROWTH AND EVOLUTION

2012
LAUNCHED MASSIVELY. BY END OF 2013 HAD 5 MILLION SUBSCRIPTIONS

2014
GREW DRAMATICALLY WITH OVER 15 MILLION SUBSCRIPTIONS AND 195 MILLION MESSAGES

2015
GOT OVER 40 MILLION SUBSCRIPTIONS

2016
EXPANDED TO MYANMAR MARKET WITH MORE THAN 2 MILLION USERS. IN VIETNAM, ZALO KEPT ON TOP WITH 60 MILLION SUBSCRIPTIONS

2017
GREW SIGNIFICANTLY BY 14% AUDEINCE IN GROWTH. LAUNCH OF ZALO PAY

2018
PUSH THE GROWTH TO 16% AND GOT ABOUT 20 MILLION MONTHLY ACTIVE USERSL

2019
BY 03/2019, REACHED TO 46,5 MILLION MONTHLY ACTIVE USERS AND ROLLED OUT E - GOVERNMENT MODEL ON ZALO IN 40 PROVINCES IN VIETNAM

2020
DUE TO COVID19 ZALO TEAM UP WITH GOVERNEMNT AGENCIES TO PROVIDE AS A COMMUNICATION MEDIUM FOR COVID10 SUPPORT RELIEF

2021
DUE TO THE ACCELERATED DIGITIZED ECONOMY, ZALO ADDING MORE FINANCIAL SERVICES, SUCH AS LOAN APPLICATIONS AND NON-LIFE INSURANCE PRODUCTS ON THEIR PLATFORM.

Zalo is a Vietnamese Super App launched originally as a messenger app in 2012. As of 2019, it has over 100 million users worldwide with more than 900 million messages, 50 million minutes of calls, and 45 million pictures delivered daily in the app. Its impact in Vietnam is hard to look over with over 80 per cent of smartphone users in Vietnam having installed Zalo on their phone. Its estimated value is $900 million. Some services that Zalo offers include Zalo Shop, Zalo Bank, Zalo Transport, Zalo Hotel, and Zalo Pay. Let's look at its other features like Zalopay, Zalo for business, and even Zalo for the government.

Zalo Pay

Zalo Pay was launched in 2017, an app through which users can use the service to pay their utility bills and other services. Users can link payment cards to make peer-to-peer payments as well as pay via Near-field communication (NFC) and Quick Response (QR) codes. Zalo Pay can also be used to purchase products and services online as well as do mobile top-ups. As Zalo Pay is integrated with Zalo itself, users can chat and send money simultaneously. It gained popularity in 2018 by running a 'lucky money' campaign during Lunar New Year where users gifted lucky money to their loved ones using the digital wallet. After the campaign in 2020, Zalo generated more than 1 million new users, 10.3 million transactions and achieved more than 2 million monthly active users in January 2020. During the pandemic, Zalo Pay had an estimated growth rate of more than 30 per cent. Zalo Pay users can link their Zalo account with multiple bank accounts.

Zalo for business

Businesses can use Zalo to sell and market their products at a lower cost and reach a wider clientele due to the high penetration rate of the app. Sellers are able to start their shop instantaneously free of charge through their Zalo account. According to a survey result of Vina research, most of the respondents answered that they have bought goods via Facebook and Zalo. Sellers can also use Zalo Ad to market their products to the large user base of Zalo. Businesses can advertise and target campaigns directly on users' feeds by filtering the audience based on demographics which include age, location, and more. Businesses are able to view the analytics on their page activity, message delivery rates, user engagement rates as well as determine the success of their campaigns. Merchants can also contact customers directly via the Zalo app to answer customers' questions, or give guidelines and explanations to customers. It is often used as a tool for customer service to gain customers' trust and loyalty. As Zalo already has a strong brand presence beforehand, this trust is already established prior which allows increased adoption of users using Zalo Shop. Users are also able to book appointments, schedule visits, and deliveries. Businesses can send and receive real-time updates from customers through the chat using broadcast messages and notifications on the Zalo Chatbot. They also need not pay for commissions to other e-commerce channels.

To improve Zalo Shop, Zalo has integrated BoxMe and Shipchung into its system. Shipchung is a popular local delivery service portal. BoxMe is a logistics and trans-ordering unit with over 3 million orders per annum. With these partnerships, merchants are now able to easily manage the system from storage, finishing orders, packing, and shipping products. This helps to reduce operating time, logistics, and costs while improving sales effectiveness and profitability. BoxMe helps merchants save up to 50 per cent on logistic costs.

Zalo for government
The connection of an online public service portal and one-stop-shop service was directed by the People's Committee of Lang Son by applying Zalo for this integration. It has enhanced the

government's efficiency and made administrative procedures more convenient for locals. Zalo account also allows individuals and organizations to submit requests online and track the progress of the request for more than 1,000 public services and seventeen departments and sectors. Local netizens are able to access the news from the government's Zalo account and update the progress of the requests each minute through Zalo messages. This has saved the time and efforts of locals to go to the public office to track the results compared to the previous way of doing it. The head of the IT office under the Department of Information and Communications said that using the application has helped save time and travel costs for locals and made the process much easier and more convenient in dealing with administrative procedures. This is a great example of a Public–Private Partnership (PPP), that paved the way for government-to-person payment(G2P) and disbursing of grants, especially during the pandemic period. We would see more of these examples when we discuss tech as a force of good in the next chapter.

Super App Lifestyle

In conclusion, Super Apps are popular in the Asia-pacific region not only because of the convenience that it brings but also the environment that fosters the growth of Super Apps, which is a huge factor for the success in the East and not the West. With government initiatives and founders working relentlessly to take the apps to greater heights, there is more competition in the race to become the first or next Super App in the country. Even though the Super Apps discussed in this chapter may sound similar, it is created for the very market they are serving for. It is important to note that each country in the region has different cultures, economies, and commercial practices. Hence, even when the Super App offerings are similar, they are tweaked to suit the end consumer in mind, through localization which would determine the success of the app.

Super Apps such as GoTo find it difficult to expand outside of the market as their unique products and services do not really fit with other cultures. So, it is more likely that we will see a Super App for almost every country. However, is it possible for the Super Apps to ensure each of its users is financially included?

We see how China really gets the 'idea' of Super Apps and has been successful at implementing it compared to India where the idea is still coming about.

'Do you think it's because the base is quite fragmented or the stickiness to one app?'

According to Sameer Nigam—CEO, PhonePe— Tencent and Alipay are some of the very successful Super Apps from China and it's because China deals with a larger scale. Well, what do we mean by larger scale? China has already been widely and massively using card payments and for a Super App to be successful in China is quite easy because it's 'a channel switch from offline to digital'. The idea of going completely digital for everything is easier to implement as compared to India, where we have just started but are picking up the pace quite fast.

Chapter 5

Tech for Good

We already know the major differences between the East and the West. We know the factors that led to the domination of Super Apps in the East. Now let's dig in deeper on 'why Super Apps are the future?' What makes an idea great or 'futuristic'—solves some long-standing problems, help save lives, and can be monetized? Or is it the underlying technology that does the trick?

We will explore how tech can be an enabler and a force of good.

A Force to be Reckoned With

The internet disrupted hospitality through Airbnb, the internet disrupted transport through Uber, and tech giants tango with the financial world is disrupting and democratizing financial services. With COP26 discussions taking centre stage, sustainability isn't merely a buzzword. Achieving sustainability and making finance work for everyone is the need of the hour. To meet the unmet needs of the unbanked and underbanked is where Supper Apps can really play a crucial role. Do you agree that Super Apps solve big social and environmental challenges?

'During the pandemic we were able to work with insurers and put out very, very quickly, very affordable products with the right level of coverage,' says Sameer Nigam in an interview which brings light to how companies like PhonePe came forward to do justice to the idea of 'Tech for Good'.

One way could be through financial services. We typically think of Super Apps as a one-stop service centre for all our needs, and financial services have become a must-have for most Super Apps. With the advent of digitalization with more Super Apps diversifying their services within the financial sector. But, can Super Apps become a tool to promote financial inclusion by delivering in a responsible and sustainable way? Let's find out.

Are Super Apps an Equalizer?

Though financial inclusion has been on a rise, financial measures used to measure financial inclusion such as income still display much disparity. There are an estimated 1.7 billion adults who are unbanked as of the World Bank's latest report. Almost half of these people come from Asian economies. Unbanked people may usually include refugees, immigrants, and cash-dependent communities. Unbanked people may be shunned away by traditional banking systems as they are considered to be risky based on looking at their past transactions and their lack of funds is considered a turn-off for traditional financial institutions. This has led to a lack of financial inclusion especially in the Asia-Pacific region, with more than one billion people not having access to formal financial services, which has become a pressing issue that is in dire need to be solved. The banked population in the Asia-Pacific region has increased from 65 per cent to 76 per cent in 2020. However, there was no growth between 2019 and 2020. To understand Super App's impact on financial inclusion, it is important to understand the three stages of financial inclusion.

First stage: Access

The first stage of financial inclusion is to enhance basic access to financial services. Access to a transaction account allows to store money and send and receive payments. Individuals and businesses should have access to services such as transactions, payments, savings, credit, and insurance. As accessing most financial services requires a significant amount of capital, it limits the reach of the service. Once access to financial services is established, it is now time to move to the second stage.

Second stage: Usage

The second stage would be increasing the rate of utilization, moving from access to the account to the usage. While there may be basic access to financial services, financial products need to be useful and affordable to ensure their utilization. Not only should the products be accessible, but they should also address the needs of these individuals. For instance in India, FinTech innovation enabled mobile/digital payments and the government's push to open zero-balance accounts has led to disrupting the financial inclusion landscape.

Third stage: Quality

The quality of the financial products and the service delivery is the last stage to make it affordable while adding value. Too many fees will prevent people from continuing to use the account. Hence, it is important to manage the types of fees and ensure that these fees are reasonable. Institutions must ensure that their value of products is realized by people. This can be done by educating the public on the benefits of different types of financial services and products.

As more Super Apps are developed for each respective country, it makes the Super App more personalized to the users. An example would be Grab. Unlike its rival, Uber, which continued with its global strategy and non-cash payment method, Grab had made every effort from the beginning to deliver hyper-local, courteous, and accommodating service, right down to the transportation they offered and accepted localized payment method, which includes cash. Typically, when a Super App is developed, it considers the cultural sensitivities, behavioural traits, and preferences of users in the country that they are looking to operate in. The process for this may be ongoing with the operation of Super Apps but these apps are created to bring ease into the lives of people it is targeting.

Super Apps that are seeking to offer financial services often target the large market of unbanked people. The lack of big players like traditional financial institutions makes the barrier of entry low for aspirant Super Apps to enter. Despite the relatively new technology of Super Apps, this does not discount its competitive edge against other firms. The edge that provides Super Apps over traditional firms would be its bountiful amount of data. The collected data from users allows them to improve the user experience and the operational processes. This data also allows Super Apps to have a better insight into its customers. Big data allows for customization and the use of Artificial Intelligence (AI) enables Super Apps to be more agile and customer-centric with personalization in the offerings. Super Apps can infer from its users' behavioural patterns in the app to develop suitable financial products for its users. Types of data that may be used to assess the risk level of loan applicants include social media and transactional data. Traditional firms do not have the same infrastructure as Super Apps. Thus, they are not able to gain the same level of insights as Super Apps. It is also important to note that the Super App's data is more valuable compared to traditional firms due to the high amount of time spent on the app. It can be seen from the case of WeChat where its users spend 95 per cent of their mobile time on it, which can be equivalent to 432.6 million hours on average over ninety days.

We've talked about financial inclusion which, to be honest economically speaking, is quite an important thing, isn't it? It's good socially too, making finance work for everyone but at the end of the day why would someone have a stake in it? How is it benefiting the stakeholders? The next section of this chapter talks exactly about this.

Advantages of Super Apps for Relevant Stakeholders

Financial institutions

Banks may be affected by the rise of Super Apps as it can possibly lead to customer attrition. Customer attrition refers to the loss of customers by a firm. As Super Apps have vast amounts of data available on hand, they are able to provide personalized services to their users. This key benefit is not offered by banks. Hence, Super Apps may seem like a more enticing option to customers. Banks being a legacy institution are heavily regulated and they cannot move at the pace of start-ups. Tweaking or launching a new online channel experience would be a long-drawn process, compared to an app that is more start-up-like and agile. However, with the use of Open Banking (a framework that allows third-party to access financial data through the use of application programming interfaces) the consumers can access their bank records, make payments, and perform other traditional banking operations from Super App eliminating the need to access the bank's app and in turn, the bank can do away with creating its own app by partnering with Super Apps directly.

Businesses

Small and medium scale enterprises (SMEs) also stand to gain benefits of Super Apps as well. The growth of SMEs can increase economic growth which helps to improve the economy as a whole. The growth of SMEs is one of the vital points for the economic development of a country. Not only does it accelerate economic growth, but it also creates employment opportunities for citizens of the country. The social development created by SMEs comes from the creation of jobs, reducing income inequality, and improving the skills of workers for better productivity. According to the World Bank, SMEs make up an estimate of more than 50 per cent of employment and 90 per cent of businesses worldwide. With the ever-growing population count of the world, more jobs need to be created to mitigate the impact of the increasing population. Hence, most countries are focused on developing their enterprises. However, SMEs need the right environment to survive. An environment that has public policies and financial services readily accessible to them. In the same article by the World Bank, it is said that access to finance is the second most cited as the biggest barrier to further develop their businesses. Super Apps are solving the access to the capital gap.

Grab Financial Group offers a variety of lending and insurance services in South-East Asia and has recently launched Grow with Grab, a bouquet of services for SME owners, they can borrow up to $100,000 through an application online. The loan has a low-interest rate per month starting from 0.7 per cent. Not only that, owners do not need to pay a security deposit. It is hard to miss repayments as the owners are notified of the repayment schedule on their dashboard.

Super Apps are like a buffet offering, out of which financial services is the most suited to help with financial inclusion.

'We've captured half of the smartphone market in India until every Indian has a smartphone or a way by which you can serve them by phones,' says Sameer, when questioned about their thoughts on where PhonePe has reached in terms of financial inclusion which is actually very true and an important thing to take note of, a digital payment app or as a matter of fact any application that is digitally operated, even Super Apps has the potential to reach every household that has access to digital services, what we further need to focus on even before talking about financial inclusion is the digital reach. Digital Financial services have become a precursor to universal financial access—mobile applications offer remote populations, often untapped by traditional financial means, an easy access to broad range of financial services.

PhonePe

The emphasis in today's world is on comfort and ease. Carrying currency has taken a back seat; the actual wallet is empty but the mobile wallet is full. Mobile apps are preferred for utility bills, phone recharges, groceries shopping, sending money to friends, splitting the money with friends after a meal, etc. Even large purchases are now made digitally instead of in cash.

PhonePe is one such payment app that has made the lives of millions of Indians easier. Users in India can use the app to link their credit and debit cards to a mobile wallet and make digital payments.

PhonePe was founded in December 2015, it is the first payment app built on the Unified Payments Interface (UPI). Flipkart acquired PhonePe in 2016.

During the pandemic, PhonePe was the leader in the UPI-based transaction category, growing at a rate of more than 800 per cent. PhonePe surpassed its competitor, Google Pay, in terms of market share. The company offers a suite of financial services such as tax-saving funds, digital gold, and insurance products (car, bike, domestic travel insurance, dengue, malaria insurance, personal accident cover, and international travel insurance). PhonePe also has an agreement with Aditya Birla Sun Life Mutual Funds that enables users to invest in 'liquid funds,' allowing them to earn returns on savings sitting idle in their wallet accounts.

In January 2020, PhonePe launched international travel insurance in collaboration with Bajaj Allianz General Insurance. PhonePe was India's first digital payment platform to do so. PhonePe became the fastest-growing insurance-tech distributor in India, selling 500,000 policies in the first six months of 2020. Acing the digital gold business by capturing a 35 per cent share. Mutual funds, which were not that big a category for the company till the beginning of 2020, have grown to over ₹100 crore in assets under management.

Where PhonePe has rolled out mutual funds, insurance, and gold investments. With over 700 apps including big names such as Ola, Swiggy, redBus, Goibibo, Myntra, Delhi Metro. The six-year-old payments company, however, lacks complimentary services such as streaming and hyperlocal delivery, which could help it evolve into a Super App.

The company is determined towards its vision of financial inclusion within India. They aim to have 500 million users by the end of 2022, they have set future plans for adding features that would deliver value to their customers and help them simplify their life. In 2020, the Mojaloop foundation appointed PhonePe to its board as Sponsor members. The members are to give technical guidance, governance, and strategy to strengthen the Mojaloop foundation. Mojaloop is open-source software that creates digital payment systems to increase financial inclusion based in Africa. In December 2020, PhonePe reported that it has processed more than 11 million insurance premium payments with more than half coming from users in tier-2 and tier-3 cities.

Grab

Around 2.8 million drivers, multiple on-demand transportations, bill payment, food delivery, and hotel booking all taking place through GrabPay. GrabPay is a mobile wallet app and payment service that is positioned as a more affordable, inclusive, and convenient service in the SEA market that has extremely low bank penetration rates. GrabPay is regulated by the Monetary Authority of Singapore (MAS) in Singapore. GrabPay uses seamless in-app technology to enable payments not only for rides but in shops and for Grab Food deliveries. Once the card details are saved on the app, users can top up funds through their prepaid account. Users earn rewards at GrabPay merchants, encouraging them to stay loyal to brand Grab.

GrabInvest is part of Grab's financial services that automates and simplifies investing for Grab users. One of its popular features is the Autoinvest. Auto invest is linked to the users' GrabPay wallet where every time users make a transaction on the app; they can choose the denomination as low as $1 in which the funds will go into their investment account within the app. The money in the fund is flexible and users can transfer it to their bank account within two to four days. With no lock-in period, this makes investment less daunting for those that are not aware of investments. The investments offer stable returns as the portfolio consists of money market and short-term fixed-income mutual funds with 1.18 per cent per year. Users can be assured that their investments are managed well with asset managers from reputable organizations such as Fullerton Fund Management and UOB Asset Management managing the funds. Grab is compliant with the Monetary Authority of Singapore (MAS) regulation, giving its users much-needed confidence. AutoInvest is a great step towards financial inclusion as it encourages more people to start investing based on the funds that they have and it has been made simple. Investment is often viewed as something that is daunting and hard to learn. With AutoInvest, Grab allows users to start investing in just a few easy steps and with no minimum investment requirement, the investment space is finally open to migrant workers, domestic workers, and people who have little to no knowledge about investing and very little investment corpus.

Meanwhile, Shop Now Pay Later also known as Buy Now Pay Later (BNPL) is an initiative for Grab Drivers. It offers discounts and BNPL deals on smartphones, groceries, electronics, and more. There are no late fees and drivers need to pay a weekly installment totalling the entire costs of products. Drivers do not need to worry about missing the deadline to pay their bills as the costs will be automatically deducted from their cash wallets on the app. This is great for people that are looking to become a Grab driver but lack the funds to get proper smartphones and electronics for their job. To protect its interests, Grab will ensure that the bills are paid through various means such as deducting from the credit wallet every day of the week. If the bills have not been paid for two weeks, Grab has the authority to make the smartphone unusable until the driver has paid for their due fees. As Grab Drivers are fundamentally important to Grab's ecosystem of services. To ensure that there is still a steady stream of drivers in the ecosystems, incentives like this should be given.

Grab has partnered with Singapore's National Trade Union Congress (NTUC) Income to launch the first micro-insurance for critical illness protection for Grab drivers. It is a flexible pay-per-trip plan where drivers can choose to pay from $0.10 to $0.50 per trip which will be deducted from their cash wallet in the app. The insurance can accumulate up to $200,000 for 360 days, which will still be in effect even if the driver is taking a break from driving until the policy expires. Drivers also need not complete a minimum number of trips to be eligible for the insurance. However, the number of trips needed for maximum protection is dependent on the driver's age. The older the driver is, the more trips they will have to make. Critical illness protection is important in Singapore as life expectancy increases. According to the Life Insurance Association in Singapore, the Protection Gap's study shows that only 20 per cent of the economically active population are covered for critical illnesses.

This micro-insurance addresses this issue and encourages more to take up insurance for their critical illnesses.

Recently, MAS issued digital banking licenses in Singapore. Such digital banks will cater to unbanked consumers and small businesses in Singapore. Digital banks have the ability to reach a bigger segment of consumers, especially the last mile. Such banks have a cost-efficient model and low entry barriers that make them affordable and accessible to underserved market segments, especially SMEs. Digital banks' offerings are similar to traditional banks minus the physical infrastructure and they potentially automate some of the typical banking operations. The full digital banking license is granted in stages. There are two stages that consist of the entry point and then the grant of the license. The first stage is the Restricted Digital Full Bank herein the bank can only offer simple credit and investment products. The MAS will restrict the number of deposits allowed, and they will also be barred from offering complex investment products such as structured notes and proprietary trading. They are also only able to operate in no more than two overseas markets. The aggregate deposit cap would be $50 million which can only accept deposits from a limited range of depositors. Each individual depositor cap is $75,000. The benefit for Stage 1 banks would be that the lower minimum capital requirement levied would be only at $15 million. Banks can then move on to Stage 2 if they deliver on the value propositions required and can manage the risks. Once the MAS has confirmed that the banks do not pose risks and are following the requirement, the bank is offering a full digital banking license that has no deposit cap and our required to have a minimum paid-up capital of $1.5 billion. The banks, regardless of stages, should be compliant with the unsecured credit rules.

Grab partnered with Singtel, a leading telecommunications company, to form a consortium for the full digital banking license. The competition for this license was tough, as there are only four digital banking licenses given out. In this entity, Grab has a 60 per cent stake while Singtel has a 40 per cent stake. Its consortium aims to address the unmet needs of underserved consumers in Singapore such as gig workers with flexible incomes and SMBs that usually have limited access to financing.

Such bank accounts with no minimum deposit and caters to digital-first consumers, which are the first and second stages of financial inclusion. With both Grab and Singtel having dabbled with financial services, as well as their services complementing each other, they aim to create a new, digital-first model of banking that is convenient, affordable, and easy to use, which is the second and third stages of financial inclusion. Singtel has been innovating its telco business to gear up for its digital bank avatar. Telcos are capable of offering last-mile financial connectivity, a model seen worldwide, from Airtel to MPesa. It remains to be seen what the future holds for Singtel's 700 million customer base.

Singapore is merely a launchpad for expansion into countries around the ASEAN region. The digital banks in Singapore very soon may target markets such as Indonesia, an attractive market for such banks. With its digital economy likely to hit $124 billion by 2030, 42 million underbanked, and close to 92 million unbanked adult population. It is the low-hanging fruit digital banks are looking to serve. The underserved SMEs will be assessed based on alternative credit scoring and unbanked people can have deposit accounts without any minimum balance requirement.

PayTM

India's much-celebrated unicorn Paytm is the brainchild of Vijay Shekhar Sharma. He set up One97 Communications Limited, which owns Paytm. This Supper App has garnered a sweet spot with investors, stalwarts such as SAIF Partners, Berkshire Hathaway, Softbank, Ant Financial, to name a few.

Paytm provides customers, offline merchants, and internet platforms with full-stack payments and banking solutions. Banking, investments, financial services, and payments are all part of the company's objective to integrate half a billion Indians into the mainstream economy.

The recent acquisition of insurance provider Raheja QBE by FinTech giant Paytm for $74 million heralds a metamorphosis in the Indian financial services market, where penetration of financial services and offerings remains low despite rapid digitization. Paytm's fundamental objective has always been to bring millions of underprivileged Indians into the mainstream economy, ever since the FinTech firm and its astute Founder and Chief Executive Officer Vijay Shekhar Sharma grasped the reins of India's financial revolution over a decade ago.

India has emerged as a leader in widespread adoption of digital payments, expanding from an online-only paradigm to complete 360-degree adoption of payments via mobile phones, mostly through wallets, Quick Response (QR) codes, United Payments Interface (UPI), and cards, over the last several years. Despite this, insurance products have one of the lowest penetrations rates among financial services. Inaccessibility to financial institutions, especially for people living in rural and remote locations, is a primary cause for the low penetration of formal financial services such as insurance, credit, and savings.

FinTech, on the other hand, has turned this on its head by not only putting financial services in the hands of people everywhere but also assisting them in signing up for such services using independent identity verification services and allowing them to tailor financial products to their specific needs. Paytm intends to provide a range of financial services including savings, credit, protection, and wealth management, to 190 million Indians who are unbanked. After China, India is the world's second-largest nation in terms of the unbanked population. The recent acquisition of Raheja QBE is Paytm's endeavour to diversify its portfolio and make insurance accessible for everyone.

Paytm has over 100 million Unified Payments Interface (UPI) transactions, 220 million stored cards, 300 million wallets, and about 60 million bank accounts, making it one of the most complete digital banking systems in the world's biggest democracy. A trip to a remote village in India shows how PaytmKaro(Pay by Paytm) is synonymous with 'cash do'(Pay in Cash). According to industry analysts, its competitive advantage of a large and growing customer base used to purchase financial products online could help it outperform larger rivals like Coverfox and PolicyBazaar, as well as withstand challenges from traditional players who are increasingly going online. For the financial year of 2020, Paytm Payment services processed more than 4 billion transactions worth ₹4.6 lakh crores, issued 58 million digital debit cards, and has a debit cardholder in every district in India. Grab in Southeast Asia, has partnered with governments to help with government-to-person (G2P) payments. This way the government can reach out to those who don't have a bank account but an e-wallet. A partnership like that may also help in developing markets in South Asia to help economically disadvantaged people, especially if they don't have a bank account or lack the awareness of these subsidies.

Not only do businesses need easier access to funding, but there should also be more ways where businesses can expand their reach and services. As e-commerce and digital services are booming, it is important that businesses are updated on the trends and are able to act upon them for maximum profit. With this, Paytm has various financial features and services catered to businesses. Businesses stand to benefit from implementing Paytm as part of the payment methods accepted. For example, businesses can accept payments directly to their bank account through UPI. Customers also benefit from this as they can choose their favourite UPI for payment. Through this integration of Payment Gateway and UPI systems, Paytm has more than 90 per cent success rates in transactions. Businesses are also able to view insights with the dashboard. They are also able to manage the account and view payments, refunds, reports, bank settlements, and more. Some businesses that are using Paytm include Zomato, Uber, and OYO. There are three types of UPI solutions offered by Paytm, which are UPI Intent, UPI InApp, and UPI Collect. UPI Intent allows customers to pay via any UPI app. UPI allows customers to pay via a bank account linked with the app. UPI Collect is where businesses can collect payments without separate integrations of services. Businesses are able to implement all three services onto their platform.

Paytm recently witnessed a 50 per cent increase in general transaction value, 121 per cent increase in offline merchants' transactions and a more than 200 per cent rise in payments for streaming services. Despite all this, it's yet to make any profit. In the capital market, it's still infamously called a 'cash guzzler', leaving people puzzled about its business model leading to a not so heroic entry at the box office alias the IPO at the Bombay Stock Exchange. But the founder claims to know what he's doing and believes that Paytm offerings are one of its kind in India. It has helped businesses to stay afloat during the pandemic with their services such as Android all-in-one POS machines for contactless payments and allowing businesses to access quick loans to avoid seeking dangerous loans with predatory rates. In November 2020, it disbursed loans to Micro, Small & Medium Enterprises (MSME), valued totaling ₹1,000 crores by March 2021.

Additionally, in March of 2021, Paytm took it further by doing a social experiment in collaboration with Dentsu Impact (a digital marketing firm) where thirty people (both men and women) from different walks of life were quizzed asked a range of questions from 'If they learnt to ride a bicycle before the age of ten?' to 'Do they handle their own finances and earn without assistance?'

and 'If they know their exact pay breakup?'. Participants were instructed to take one step forward if the response was YES and one step back if the answer was NO for each question. Men took steps forward as soon as the questions touched on personal finance and financial freedom. The financial gap was startling between men and women. That is when Paytm found out that there is a lack of financial awareness among women. The women who participated in the experiment were educated and had jobs such as writers, directors, and strategists. They were disappointed in themselves for not knowing certain terms and some shared that they were never given the opportunity to learn more about finance as the importance was given to their male counterparts in their family or at work. This social experiment was strong evidence of the huge gap in financial literacy between men and women. Paytm has taken actions to improve financial literacy such as conducting masterclasses designed for women to learn more about investing. Paytm Money presented 'The Women Investors Masterclass, exclusively for women investors', where attendees learn from some of India's most remarkable female investors as they share their personal journeys to financial independence, understand the difference between investing for short-term and long-term to plan their financial goals sensibly and much more.

Koinworks

Koinworks is an Indonesian peer-to-peer lending app that aims to be a Super App in financial services. Koinworks is doing well with 80 per cent of users wanting to repeat the loan application. Started in 2016, Koinworks offers peer-to-peer (P2P) lending that has promoted financial inclusion within Indonesia. It has also made SME online financing more accessible.

Despite their significant contribution to the economy, the SMEs market in Indonesia has not been very utilized by the banking industry as there is a huge funding gap. The data published by Bank Indonesia and the Financial Services Authority show that in 2018, only 19 per cent of the loans given by the banking industry were allocated to SMEs. The total value of the loans given was approximately 5,168.2 trillion rupiahs. However, the funding gap was about 1,320 trillion rupiahs. To address this issue, Koinworks has brought several features that help to address the needs of SMEs.

Koinworks' peer-to-peer lending model has a proven advantage over the bank model which is their lack of interest margin. It currently has more than 500,000 retail lenders and 4,000 borrowers on the platform.

Without the interest margins, both parties in the peer lending model will benefit as there are lower interest rates for the borrower and lenders can gain more out of the deal. There are several products offered by Koinworks such as KoinBisnis for SME financing and KoinPintar for financing education. Their education loans are a great way to improve economic growth and human development as there will be more skilled workers in the labour market.

With their data on SMEs collected from various sources such as e-commerce, POS, and other software, they are able to assess the potential borrowers and approve them quicker. It uses technologies to analyze fraud and risk-based on-site behaviour, multi-source verification, credit scoring, and revenue verification. As the COVID-19 pandemic hit, Koinworks continued to be profitable as it pivoted to a different strategy to address the needs of SMEs. As more businesses are digitally connected, Koinworks was able to take advantage of this as they are now able to serve more people.

One of its key features for merchants would be its pre-approved loan. These merchants are selected based on their performances in the partner's platform and are offered loans with same-day approval. This helps to increase loan approval rates and at the same time, convince lenders to use the service as these borrowers are pre-screened before selection. Partners of Koinworks include Tokopedia (e-commerce company), Lazada (e-commerce), MokaPOS and JNE. MokaPOS is a point-of-sale software for businesses and Jalur Nugraha Ekakurir, better known as JNE, is an express and logistic courier service. Data is collected from these platforms for analysis of prospective borrowers based on their order fulfilment rate, delivery rate, and other financial measures that may be used to determine the risk of the borrower.

WeChat

China is the world's second-largest economy but it has the world's largest unbanked population: with 225 million Chinese adults currently lacking a bank account. WeChat's parent company, Tencent, launched WeBank in 2014 to bridge the financial gap. WeBank offers microloans that can be applied directly from WeChat Pay. The credit scoring is based on the user's transaction. Thus, people who lack financial or credit history stand to gain. The smartphone adoption rate rose from 39 per cent in 2013 to 71 per cent in 2016, e-commerce wave motivated people to create digital banking accounts to purchase or to become a merchant. The platform now serves more than 100 million customers that were previously underserved.

WeBank's first loan product was an unsecured loan offered through WeChat and QQ's wallets which has now become an accessible way of financing for China. Furthermore, more than 70 per cent of the borrowers are blue-collar with an average amount of more than $1,000. Application for credit approval is almost instant where they can receive the application result and funds almost instantaneously. WeBank has also been financing more SMEs where two-thirds of the borrowers are receiving a loan from a financial institution for the first time. Its banking costs are three to six times lower than traditional banking operation costs, where its cost to serve per customer is about $0.50.

This means that WeBank is able to streamline its processes and cut out unnecessary fees that may inflate the cost of repaying the loans which can deter consumers from seeking financial help in proper channels. It is able to keep the costs low with its investments in technology. WeBank focuses on AI, Blockchain, Cloud Computing, and Big Data for its processes. Most customer inquiries, about 98 per cent, are resolved by WeBank's chatbot which then reduces the need for dedicated staff to attend to the inquiries instead. It has ensured the security of its services, where over 600 million identity verification requests are fulfilled through WeBank's facial recognition software.

There are about 200 million International WeChat users including 19 million daily active users in the United States. At first glance, the number of international users seems big, but when you compare it to the total number of users, that is 1 billion, international users only make up to 20 per cent of its total users. The main reason for its success is due to the critical mass it has gained from the 1.408 billion (2019) citizens in China. With China being the world's most populous country, its domestic market is enough to make its app a success.

WeChat has definitely provided financial inclusion with its WeChat Pay as it has made it frictionless to pay for products and services this is due to its network of authorized business partners in thirteen countries.

Wave Money

Wave Money was established in 2016 as a mobile financial services provider in Myanmar. It was a joint collaboration between Telenor Group, Yoma Bank, Yoma Strategic Holdings, and First Myanmar Investment. Wave Money is successful with its transactions amounting to $4.3 billion just within 2019. Despite the pandemic, it has recorded a total transaction worth $8.7 billion. WavePay app launched by Wave Money recorded 1.6 million monthly active users. It has also launched a financial education game app to educate women on financial concepts in 2018.

It collaborated with Myanmar Economic Bank to provide mobile pension payments to government pensioners in 2019. Wave Money later went onto to gamify the financial learning process such as trading, investing, interest rate calculation, etc.

The company pivoted its model during the pandemic to partner with the government to provide grants to its migrant workers who returned home after they lost jobs in Thailand and other markets.

The app can be used for bill payment, WavePay to WavePay free transfers, and online purchases. The transactions are instant and secure with options to use QR payment as well as linking the account to their bank account. Users that have a Yoma bank account can connect it to the account on Wave Money without the hassle of visiting the bank branches. Users can upgrade their accounts from Level 1 to Level 2. Level 1 has a wallet capacity of 400,000 MMK, a daily transfer limit of 100,000 MMK, and a monthly transfer limit of 2,500,000 MMK. For Level 2, its wallet capacity is five times more than the wallet capacity in Level 1. It has ten times more capacity than level 1 for the monthly and daily transfer limit. With WavePay, users are able to top-up their mobile any time and anywhere. They are also able to pay their loans on time easily and instantly. Wave Money has partnered with MMBusTicket, Oway, Flymya, and other travel companies for users to easily purchase bus tickets, flight tickets, and hotel reservations within the app itself. Users can also purchase game items using Wave Pay with promotions, discounts as well as cashback on the transactions. Users are also able to withdraw the cash for free by going to the nearest Wave Money shop or from a mobile banking platform.

It is also the first FinTech in Myanmar that launched an open Application Programming Interface (API) Platform to enable cashless payments between merchants and customers. This allows merchants to connect their e-commerce businesses to the platform for cashless payment on goods and services from their customers. Wave Money organizes 'Wave Money Merchant Clinic' a forum that encourages dialogue between merchants and the company. The company, unfortunately, couldn't seal the deal for $3.5 million funding from Ant Group due to political instability in Myanmar.

Potential privacy concerns of Super Apps

One of the key concerns with Super Apps is that they collect huge amount of data, all this metadata could fall into the wrong hands. As seen earlier in case of different data controversies, companies now more than ever need to ensure data security and privacy. It's not new for the apps to access, store, and utilize user data but sometimes the data is misused. With the privacy concern in mind, people are hesitant and sceptical about getting onto different platforms—the what-ifs and the possibility of their data being used in the wrong direction is something that can be a hurdle for the adoption of Super Apps and its consumer base. The recent ban on Paytm bank by the Reserve Bank of India (RBI) to acquire new customers due to breach of data laws is a perfect example of data being at the centre stage of the regulator's attention.

PhonePe shares how the company differentiates itself from the competitors by choosing the amount and type of data collected from customers. They are particular about the refineries of data they want to build, currently, their motto is less (data) is more. The company acquires transaction data on a consent basis, and the data that is captured is used to offer better products to customers, such a way that consumers will get more value for their time. PhonePe also has a strict policy of not sharing data with any organization outside the company. 'When less is more' privacy issues are sure to stay at bay.

'So, we asked for her location the first time we booked a cab. We asked for your contact book if you want to send money to your friend.'

Sameer is a classic example of asking for consent before acquiring any personal information about the consumer.

Consent is a great way to take customers into confidence and build good trust and rapport while businesses discover a whole new ecosystem of personalization. Don't you think companies have a fiduciary duty to safeguard customers' personal data whilst exploring new possibilities to delight them?

'It needs to be consent-based so we are launching. In fact, lateral consent is supposed to be over.' Says Sameer Nigam while talking about data refineries.

Super Apps offer a world beyond banks, bringing electronic payments and other financial services to the last mile with efficiency and convenience. Gartner a technological research and consulting firm, expects that by 2027, more than half of the world's population will be active user of multiple Super App.[xi] The breadth and depth of user data generated and collected by Super Apps unlocks new inferences: digital and credit trails, that can prove to be a rich vein to tap, not to forget, treading the boundaries of data privacy.

Chapter 6

Temporary or Permanent?

To answer the question if the Super Apps are here to stay, it would most likely stay based on the ever-growing developments in the Super App industry, which we have touched on in previous chapters and will explore in more specific detail below. More and more companies are looking to create their own Super App but are restricted due to the high cost and other factors depending on the region that they are in.

Undoubtedly, while the adoption of Super Apps in the West may take longer, it will eventually be that there will be more Super Apps to be developed. In the chapter below, we will uncover some of those Apps. As is evident with Uber, taking strides towards becoming a Super App. The pandemic has also become an almost perfect environment for Super App to thrive, where new lifestyle changes require more tailored technology to aid in the changes. There are also more Super Apps popping up in other regions which we will be discussing soon.

Initial Public Offering

Super Apps capture mass market so inclusion of the public in the company's growth to raise capital while amassing better public image sounds like the right strategy for scaling up. The first selling of stock issued by a firm is known as an initial public offering

(IPO). In other words, it occurs when a company decides to begin selling its stock to the general public. The company will decide how many shares to offer, and an investment bank will recommend an initial price for the stocks based on anticipated demand.

Super Apps have entered the capital markets. Grab took a decade to get to reach its IPO stage. Their journey began when the founders placed second in a graduate school business competition and has now evolved to Grab operating a Super App in eight Southeast Asian nations, providing services in three categories: delivery, financial services, and mobility. Mobility has experienced rapid expansion, particularly in the delivery and financial services industries. In 2020, Grab dominated 72 per cent of the ride-hailing industry, half of the online delivery market, and nearly a quarter of the digital-wallet payment market.

It is Southeast Asia's first 'decacorn' - valued at more than $10 billion. The company's CEO has been open to an IPO since at least 2014 when he stated that an IPO would be considered whenever the number of bookings through the app hit 2 million per day. Grab recorded whopping 1.9 billion transactions in 2020. Their IPO is unique as it is held concurrently with a SPAC merger with Altimeter. Special Purpose Acquisition Company (SPAC) is an acronym for Special Purpose Acquisition Company. After the merger, CEO TAN has 60.4 per cent of the company's voting power.

One97 Communications—the company that owns Paytm IPO continues to tank on the stock market—nearly wiped 50 per cent of investors' wealth since its IPO in November 2021. The Noida-based company was initially backed by Alibaba (which owned 36 per cent of Paytm alongside Ant Financial), Berkshire Hathaway, and SoftBank (which owned around 18 per cent).

COVID-19 Impact and Possible Acceleration of Super Apps

The pandemic has accelerated the growth of the Super Apps not only in its native markets but also in other sectors, by providing innovations. GoTo (result of a merger of Gojek and Tokopedia in 2021), Indonesia's largest digital company, had to let go of some of its taxi drivers but managed to hire more drivers for their food delivery. In 2022, the company had the world's largest initial public offerings priced at IDR 338 per share it raised $1.1 billion, winning over rivals such as Grab. With the new normal far different from what our lifestyle used to be, there are now more innovations that help with the growing pains of these transitions. There is now widespread use of Super Apps like Grab for food, which has been accelerated due to the pandemic, and grocery delivery services as lockdowns reduce people's movement. With the rise in usage of online services, Super Apps are able to utilize this opportunity through reorganizing their services and determining which to add or eliminate in their scope of services. Even though Super Apps have been around before COVID-19, the pandemic has allowed the apps to be promoted from just being used as an alternative to an only option.

Potential Super Apps in healthcare

Since the pandemic, efforts in telemedicine have skyrocketed. Services like virtual care delivery and in-home patient monitoring which were expected to be widely available only in a decade, are now soaring in this pandemic. As the pandemic loosens the tight regulatory constraints and medical restrictions lead to more healthcare providers taking up medical technologies such as telemedicine. In replacement for on-premise visits, asynchronous consultations are done instead. Home care can be improved with sensors and AI such as monitoring kidney function and heart failures. Dr John Halamka from Mayo Clinic has said that the advantages of telemedicine outweigh the losses incurred from lack of personal contact. He has also said that people are able to pay less for the same outcome and convenience. To support these transitions, Super Apps that are in the making, that are discussed below are made to assist doctors and patients for better healthcare services.

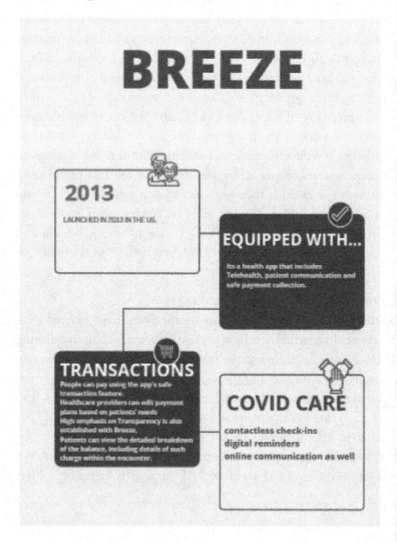

Some features of Breeze that was launched in 2013 in the US, include contactless check-ins, equipped with Telehealth, patient communication, and safe payment collection. With the pandemic, it has been highly recommended to prevent contact with high-touch items to prevent the risk of catching the virus. Hence, contactless check-in is used in many areas. With the app, patients can check-in anywhere from waiting rooms or even from their home. They can also avoid the use of tablets and clipboards just to check-in for their appointment. Through Breeze, patients are able to request, check-in and join the telehealth appointment. They are also able to get reminders about the appointment without using the staff's time to send messages. The messenger feature allows health institutions to communicate with their patients. This is especially important in the COVID-19 era as there are constant changes to lockdown measures in each region.

Patients can also pay using the app on the go for safe transactions. Healthcare providers are also able to edit the payment plans based on patients' needs with the payment plan builder. As more patients are claiming the expenses with their insurance, it is important to ensure that the claims are not denied. Breeze helps to reduce these claim denials by collecting accurate demographic information to allow eligibility verification. The process of transcription is also streamlined where it prevents errors, with faster reimbursement, and better financial performance. With these features implemented on the app, Breeze has increased the patient collection rate by 5 per cent to 10 per cent. Patients are also able to review and update their particulars and insurance details. CareCloud, the company that developed Breeze, claims that ten to fifteen minutes of staff time is saved for each patient. Transparency is also established with Breeze, as patients are able to view the detailed breakdown of the balance, including details of each charge within the encounter. Not only does this improve staff productivity as there is less need to clarify confusion or resolve complaints, but there is also stronger trust built between the healthcare provider and patients. This means that with this trust, there will be more repeat patients and they will consider the healthcare provider as their first choice.

Compliance can often be a pain for most healthcare providers. With Breeze, healthcare providers can configure consent forms based on visit type, location, or provider without having to build pre-registration packs. These users can also set triggers on various frequencies such as every month, quarter, or year. These are useful for time-sensitive forms, where the required forms are only shown when the users need them. Patients' data can be kept updated through flagging patient-submitted information for staff to update in the system. Patients are also able to access information in the app such as lab results, medications, and allergies. To comply with data protection laws, users can use Breeze to configure each staff member's access and permissions for all features and settings.

Healthcare providers can also track their inventory of medical products within the app and can get alerts on the app if there are products running out. Patients can also reduce the time spent at the place by choosing to ship their products to their homes. Healthcare providers can ship these products with ease as the app provides accurate estimates of cost and time at check-out.

Lybrate

2015

LAUNCHED IN 2015 LYBRATE IS
A MOBILE HEALTHCARE APP.

more than 100,000 doctors
approx 4million downloads of
the app

**CHATBOTS IN
LYBRATE**

launched in 2016

chatbot on Facebook
Messenger to encourage
people to consult doctors
with their health matters.
People are able to get the
inquiries answered in real-
time by more than 10,000
doctors available to answer
the queries.

HEALTH PRODUCTS
STARTING RS. 549

NEW COVID CARE
PLANS

Lybrate was launched in 2015 as a mobile healthcare app that aims to solve pressing healthcare issues such as poor access and doctor-patient ratio. Its main function is to connect patients to doctors online through a video call or an appointment scheduled. It is also used as a platform to get info about medication. Within two years, it has amassed more than 100,000 doctors with 4 million downloads of the app. Though there are several competitors that have similar services to Lybrate, it has additional services to mimic a complete experience. These services include appointments, consultations and if necessary, tests and follow up. There is also a health feed included in the app, which keeps users updated with new developments in the health sector. Doctors can also provide health tips such as homemade remedies and fitness tips on the feed as well.

A survey done by Lybrate reported that more than half of the interviewees choose to self-medicate. With more people preferring to self-medicate, there is increased drug resistance and underlying medical conditions are not diagnosed until much later. Some reasons that were cited by one of the founders of Lybrate, Saurabh Arora, include lack of time, avoidance of consultation fees, and reliance on the internet. With the high cost of fees for consultation tests, more people avoid going to healthcare providers. Though there are government hospitals, the queue is long, and can often take a long time for doctors to attend to all the patients. The average consultation fees at a typical clinic would be about ₹500 to ₹1,000. With Lybrate, users only pay an average of ₹200 to ₹350. Wrongfully prescribed medications by non-experts can cause serious health problems. Not only does this develop resistance to drugs that may not be the correct cure for their health issues, but it may also cause other health problems that may be more severe in nature. Health problems may include liver damage, stroke, and many others. The easy accessibility of over-the-counter medications is another factor for people to self-medicate. Hence, Lybrate addresses these issues with its app where people can get quality advice from experts without having to rely on questionable prescriptions from others.

In 2016, Lybrate launched its chatbot on Facebook Messenger to encourage people to consult doctors about their health matters. People are able to get the inquiries answered in real-time by more than 10,000 doctors available to answer the queries. Lybrate has also added a health quiz into their chatbot. This chatbot helps to keep users anonymous which can be helpful for people to ask questions about their health matters that they were previously hesitant about.

Lybrate has GoodKart, the e-commerce aspect of their app that offers health products from local and international brands. These health products are sold at affordable prices. There are often discounts and promo codes that users can take advantage of to get better prices for their products. Lybrate also has GoodBox where users can get curated boxes of health products by experts with a starting price of ₹549. Users can choose the category of products included in the box. This includes categories such as strength, weight, immunity, diabetes, skin and hair, and children. There are also plans that users can undertake to get bundle deals. One example would be the COVID Care plan. This plan allows users to chat and consult with doctors on the app an unlimited number of times for two weeks. The doctors may prescribe medications to alleviate the symptoms of the virus. Users do not need to pay additional fees for follow-ups during the consultation period. This plan is eligible for users that want advice on prevention from COVID-19, tested positive, and need guidance on self-care at home to prevent hospitalization, as well as for individuals that think that they might have been exposed to COVID-19 but are not validated by the test.

Lybrate is not only beneficial for users but also for doctors. Doctors are able to increase their revenue with their own 'clinic' on the app. Doctors are also able to save time and manage time better. Doctors can use the platform to discuss lab reports, adjust the drug dosage, or follow-ups. Doctors can free up their appointments for cases that need to be seen offline. They are also able to expand their reach beyond the location of their practice. With the expanded reach, doctors can build a reputation online through the app such as providing quality consultations and quality advice on the health feed. They are also able to monetize all consultations from converting to unpaid messages and calls to paid interactions on Lybrate. Workflow can be better managed with Lybrate's Practice Management Software to streamline processes and save time for doctors. Some features of the software include Electronic Patient Records, Digital Prescriptions, Appointment Schedulers, and more.

Apple Health App

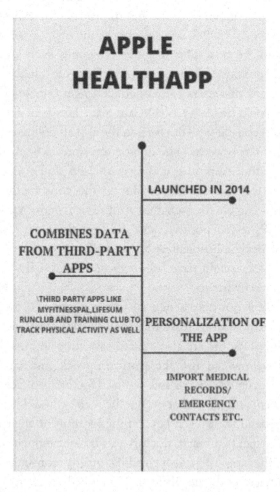

This health app was launched in 2014 with its easy and comprehensive way for users to view and manage their health data from various sources. It combines health data from third-party apps and devices such as run and sleep trackers, food diaries, reproductive health trackers, and more. The functions are split into four main categories which include activity, mindfulness, sleep, and nutrition.

The activity portion of the app shows the data of the daily activity and an overview of the entire year. It uses the device's accelerometer to measure steps and distance traveled without needing a separate tracker. To get more accurate statistics such as the time spent standing, calories burned, and more, users can buy an Apple Watch. The overview includes data from the Apple Watch such as exercise minutes, resting energy, active energy, and standing hours, and more detailed metrics such as running distance, cycling distance, and more. The Health app also supports other fitness trackers such as FitBit and Garmin. They are also able to track these types of data from third-party apps such as the Nike+ apps, the Run Club and the Training Club apps.

The app can also be used for tracking food intake but it is time-consuming as it requires users to input everything. However, third-party apps can be used such as MyFitnessPal and Lifesum, can be synced to the app. These apps already have the data required for the nutrient breakdown of meals and snacks that are entered. In the nutrition tile, one can see the breakdown of both macronutrients and micronutrients for various timelines, from daily, monthly to yearly. The app can be used for weight loss such as tracking the weight-loss progress. Users can view all the weight data from the manually entered data in the app itself, or from connected compatible apps or smart scales. The app also allows users to enter other information that provides a more accurate depiction of health. This includes data such as height, body fat percentage, body mass index, waist circumference, and more. This is more tailored for users that have different health goals. Users that want to build muscle, can look at the data for body composition rather than weight as it does not provide an accurate result. Users are also able to track their menstruation cycle with Cycle Tracking which is built into the app but can be downloaded as a standalone app. Users can keep track of their symptoms such as cramps or headaches or trackable data such as their body temperature and results from ovulation tests.

Sleep is important to one's health as it affects both physical and mental health. Users can track their sleep with the built-in feature called Bedtime on the Clock app. Users can set their bedtime and the phone will alert the users to sleep when it's time to. The phone will log in the wake-up time when the users pick up the phone. Users can use third-party apps and devices to track other data in relation to sleep. For example, users can use a Beddit device that syncs both the Apple Watch and the phone. The type of measures it tracks includes sleep time, heart rate, breathing, snoring, bedroom temperature, and humidity. Users can also use apps such as Sleep Cycle to track the quality of their sleep, time spent in sleep stages, and more. For mindfulness, there aren't many avenues to track the mindfulness aspect of the app. It is still quite heavily reliant on third-party apps and devices. The Apple Watch has a Mindful Minutes feature and there are third-party apps such as Breathe that gives reminders for users to focus on their breath and can set the frequency of reminders throughout the day. Users are also able to choose the apps that they wish to integrate into their Health app based on the suggestions that it gives.

For users that may have a history of chronic illnesses or life-threatening conditions, it is important to be medically prepared. Users can import health records from the doctor or clinic. The reports include immunizations, medications, blood pressure, allergies, test results, and more. Users can also set up a Medical ID in case where first responders need to access their health information. These responders can then access the information from the home screen without unlocking the phone. Other information that can be included in the Medical ID includes emergency contacts and phone numbers which are useful for the hospitals to contact the relevant people.

None of the three apps have mentioned if users are exposed to regulatory and medical risks, which is something crucial to take note of, since everything is online. Therefore, it is important that we ask ourselves these questions: Are these apps safe to use? Do they still need more work to be done in order to stay here for the long haul?

Productivity Super App: Lark

LARK
AN ENTERPRISE COLLABORATION PLATFORM

2019
RELEASED TO THE PUBLIC IN 2019

5
Lark has 5 main services which includes Messenger, Docs, Meetings, Calendar and Mail.

SERVICES
CREATING/EDITING/SHARING CONTENT/ VIDEO CONFERENCES/ CALENDERS.

Apart from the rise of telemedicine, the pandemic also led to increased complexity and productivity in the realm of work. Routines have become duller with a lot of tasks, goals, and endless demands for productivity. Even though working from home can be considered relaxing, the reality says different. The lines have been blurred and the home is no longer a sacred space where work is off-limits. As work has become more complex, this also results in many obstacles. Technical barriers such as access are the first to be faced when using digital devices, especially for those who live in far-flung areas with low network connectivity and incompatible devices. Another obstacle in achieving productivity is the navigation and operation of an application or the web. For workers who are not familiar with digital devices, this can decrease their productivity. Even in face-to-face classes, students can experience technical difficulties when a teacher or lecturer uses a new learning technology. The last obstacle is technical problems when conducting online activities. Unstable internet connection and errors in operating devices such as laptop audio-video components can also occur. Although it is not quite often, cyberattacks on users' privacy and personal data might also happen.

Lark Suite is an enterprise collaboration platform developed by TikTok's, BABE's, Resso's, and Moonton's parent company, ByteDance, and first released to the public in 2019. Lark has five main services which include Messenger, Docs, Meetings, Calendar, and Mail. it keeps related messages together by allowing users to reply to a message and start a thread. Users can access chat history to preserve context. They are also able to send urgent messages by sending a 'Buzz' message to ensure a quick reply from colleagues. The app also displays whether messages have been read to make communication as efficient as possible. Users are also able to pin their most frequently used contacts and group chats to Lark Messenger's QuickSwitcher. Lark also has a list of chatbots that acts as virtual assistant. Each chatbot has its own unique functions. Users can instruct bots to send reminders, post messages, and automate workflows.

Lark makes creating, editing, and sharing content the best experience you can imagine. Docs and Sheets support real-time collaboration, full history tracking, powerful, rich media, and easy permission control. Users can access their files anywhere as all the documents are saved in Drive. Users can create and edit documents with smart formatting using Docs. Users can insert images, videos, maps and more. Sheets has various features such as flexible third-party data imports, conversion to Excel as well as full editing capabilities and data representation. The Drive is optimized for teams where it allows simultaneous collaboration and multimedia insertion. Using the cloud storage in Lark Drive, files can be saved and managed any time, anywhere on any device. It also provides previews of files of various supported formats, advanced search with quick view and fast uploads and downloads of files.

Users can also link their Gmail account to Lark for integration between email, messenger and drive. All emails, labels and filters will be synced in Lark Mail without having to sign in to the Gmail account. It also sorts emails as 'Priority' automatically and allows users to organize emails using the labels and flags. In Lark Mail, users can add names of colleagues as well as chat groups which is useful to send announcements to groups. One problem that is often faced by users when sharing files for collaboration would be that it is time-consuming to check if each and every one of the recipients have access to the document. Lark Drive solves this problem by providing easier visibility to recipients' access and can easily give permission to read and edit. Security is ensured with Lark's services with the partnership of Amazon Web Services. All data which includes emails are securely stored in data centres in the US. The data are also secured through spam detection devices and phishing attacks are also prevented using the integrated web risk API from Google.

Video conferences can be started anywhere and any time with the integration of Lark Meetings with Messenger and Calendar. Collaboration can also be made using Lark Meetings with in-call screen sharing capabilities. Individual and group chats can be turned into video calls instantaneously. The integration allows users to attend the call without meeting invitations or access codes as it accesses data from Messenger and Calendar. Users can join the call from the Calendar event by clicking on the link. Local documents, Lark Docs, and the user's screen can be shared using the Lark video call. From the Lark video call interface, the Magic Share feature can be used for collaborating on a Lark Doc on the call itself. Lark Meetings can be used on both mobile and desktop devices which is great for users that are on the go.

The Calendar can be used to book a meeting, stay on schedule, and engage the team. It also allows users to subscribe to multiple calendars and display the events within a single view. Subscribing to others' calendars can be used to check the availability or check group events. Meetings' details can be easily determined by seeing each other's calendars side by side in Messenger. Using Messenger, the entire group chats can be invited to the meetings in Calendar. By adding the group's name to the event guest list, all members in the group chat will receive an invite. Users can also check the availability and book the meeting rooms in the office instantly with a single click. Chat groups can also be created from the event invite for meetings. Calendar Assistant can be used to help users automatically send event invitations and reminders.

Why Are Some Companies Not Interested in Being a Super App?

It may sound weird, in the world where many companies are rushing to be the next Super App in the region—especially more so in the aftermath of COVID-19—that there are companies that are not looking to become a Super App or to be labelled as one. One example of this would be foodpanda, an online food and grocery delivery platform brand owned by Delivery Hero, which is headquartered in Berlin, Germany. Foodpanda, today operates as the lead brand for Delivery Hero in Asia, with its headquarters in Singapore. Rather than taking action to introduce services that are not related to the core purpose, foodpanda prefers to focus on the untapped food-delivery market potential in the Asia-Pacific region. Instead, it looks into expanding customer touchpoints by developing new innovations in technology. To fulfil this, foodpanda introduced grocery delivery options, known as PandaMart, and food pickup options. PandaMart is proven successful with its outreach of 150 PandaMarts in forty cities in eight markets. Foodpanda uses technology like data analytics to see what is the number of PandaMarts needed in each city based on factors. This includes consumer density as well as the optimized location to place the PandaMarts. PandaPro offers various benefits which include members-exclusive deals and reduced delivery fees. Despite introducing the subscription model, the majority of its revenue comes from transaction fees. Hence, most of their strategies are developed with the purpose of improving user experience through product development. It is also looking to make its food-delivery app more engaging and personalized by delivering options that are best suited for its users. This may be done through data analytics using various data points and analyzing behavioural patterns. It also looks at reducing conflicts faced by users, whereby users can manage the issues themselves without the need for an agent.

Mass-producing Super Apps: Appboxo

What is it like to launch a mini-app within an app without having to worry about tech giant resources? Ask Appboxo—a Singapore-based start-up that is looking to make mini-apps more accessible by allowing developers to turn their app into a 'Super App'. It currently works with about ten apps, which include

Klook, Zalora, and Booking.com. The start-up has also partnered with banks, telcos, and mobile wallets to turn their apps into Super Apps. It also has about eighty mini-apps on its platform. Appboxo SDK (software development kit) can be integrated with an app to launch dozens of mini-apps inside the app itself. The mini-app is then loaded almost immediately which improves user experience and core business metrics. Appboxo also provides customizable features that allow businesses to ensure that their app follows its design guidelines. Apps can also connect to Appboxo Connect API (application programming interface) to prefill forms and pass information to enable one-click payments. Appboxo streamlines the process of creating the 'Super App' where businesses can skip the development of the front-end of the apps. This is useful for developers that do not have enough resources to develop their own mini-apps. When the developer installs the SDK (collection of software development tools in one installable package), they would only need three days for technical work. The installation of the SDK itself is free. But when a user completes a transaction on the mini app platform, the Super App and Appboxo will take a commission split. They also do not need to focus heavily on legal discussions. This is because its services have already defined its terms and commission rates and businesses do not need to look into commercial and legal negotiations. Looking at the changes in the 'Super App' market by Uber with its merger of its ride-hailing and food delivery service into one app, Appboxo predicts that the use of apps with one core offering will change with more users using Super Apps. Appboxo also has partners in Europe.

Developments of Super Apps

To illustrate the growing popularity of Super Apps, we will be looking into developments of Super Apps that cater to the specific needs of the market they are serving.

KiwiGo

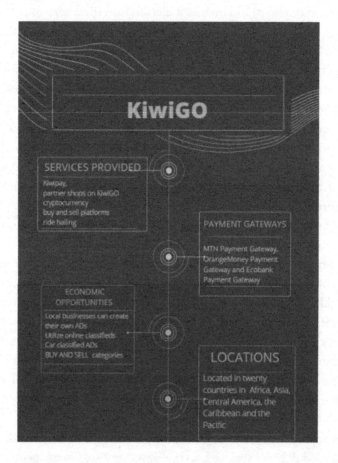

KiwiGo is a Super App for the frontier markets based in Singapore, that is present in more than twenty countries with various services such as wallet, last-mile logistics, ride-hailing, shopping, buy-and-sell features, and cryptocurrency. KiwiGo has a payment service provider called KiwiPay which is trusted by most of the bigger companies in Southeast Asia. Businesses can use this service to give their customers various payment options that they may prefer to have higher business opportunities. It also enables transactions to be made between payees and recipients from different countries. KiwiPay also includes features such as an online wallet, international remittance, and online payment gateways.

Users can buy their groceries and other products from various partner shops on KiwiGo. It also plans to add a cashback system for users. To encourage more merchants and drivers to come onboard, it is also planning to have KGO referral rewards. By 2023, it plans to add logistic tracking and Smart contracts using blockchain. KiwiGo has also integrated MTN Payment Gateway, OrangeMoney Payment Gateway, and Ecobank Payment Gateway into their own payment gateway in May 2021. KiwiPay in Madagascar has partnered with the Bank of Africa for the purpose of digitalization and to create a banking union between African emerging countries.

Users can pay with the in-app cryptocurrency, $KGO, which can be used in more than twenty countries. KiwiGo also offers ride-hailing services and taxis with thousands of drivers in Asia and Africa available on the app. There are also more options for users to choose from taxis to motorbikes. They have also listed their cryptocurrency on various platforms, which include Cointiger and Binance. As of 20 May 2021, there are more than 60,000 KGO transactions on the Binance Smart Chain.

To promote economic opportunities for local businesses, users in KiwiGo are able to create their own ads for free to be shown in the city that they wish to promote. Online classifieds have shown to be a cost-effective and efficient means of reaching a broader audience. Companies can utilize online classifieds to reach a large audience that is spread out over the country. It has become easier to market because businesses no longer have to bear the costs of publishing classified ads or having them published in local media, which are significantly more expensive. They have also integrated car classified ads integration for ninety countries in August 2021. This includes Asia, Central America, Caribbean countries, and African countries. This means that all classified ads published on ECV's platforms are now synchronized and automatically published under the automotive tags of the KiwiGo app 'Buy & Sell' category.

To Stay or to Go

The Super Apps are definitely here to stay with more and more Super Apps being created at the moment. With it now slowly expanding to the West, we are yet to see how it would change the way the entire world works. Super Apps are known to be versatile with their large number of services that can be hosted within a single ecosystem. Therefore, it is no surprise that the concept of Super Apps has become more prevalent in other sectors than finance and healthcare. We can expect more markets to adopt Super Apps to be used in daily life with seamless integration of services between online and offline.

With the continuous evolution of technology, we can only wait to see how Super Apps will further change the world in the future, hopefully with a positive impact. With how helpful Super Apps have been during the pandemic, we should remain optimistic about Super App's future.

Part II

We have covered a lot of ground in Part I of this book, exploring the history and evolution of Super Apps and their impact on the financial world. However, there is much more to this industry than meets the eye. As I delved deeper into my research, I began to realize that to truly understand the nuances of this market, we needed to hear from the people who have firsthand experience with it.

That's why I'm excited to introduce the next section of this book, where I will be featuring interviews with a wide range of experts, including bankers, venture capitalists, founders, and more. These individuals have a wealth of knowledge and insights to share about the world of Super Apps, and we can learn a great deal from their experiences and perspectives.

For instance, Bakhrom Ibragimov, a Venture Capitalist, shared his thoughts on engagement and cross-selling across verticals. Governor Serey Chea of the National Bank of Cambodia shared that regulators fear the growth of Super Apps with the large data play. She told us about how Super Apps are like the accelerators, and regulators are the brakes in a car to ensure the checks and balances in the market. Next, Venture Capitalist Keegan Sard gave us his vision for the future of the Super App industry, speaking at length about the space in Europe, Australia, and in the East. Fintech Influencer Chris Skinner offered unique perspectives on the cultural aspects of regulation, giving us insights on consumer behaviour and trends, and Professor Patrick Pun discussed decentralization and centralization of Super Apps, while giving us a consumer's perspective at the same time. Janet, MD and Head, Group Channels & Digitalization, UOB Ltd. discussed the rise of Super Apps in the post-pandemic era and how banks play along with the big fintechs. Phanindra Sama, the founder of redBus (an Indian online bus ticketing platform), shared his experiences of Super Apps in the Indian market, talking about how he built redBus, and the relevance of Super Apps in the future.

Through these conversations, this section illustrates perspectives and insights from key players and industry leaders. By gaining access to the thoughts and insights of these prominent figures, we were able to gain a more complete understanding of the current state of the industry and the potential future of Super Apps.

Throughout this section, we will be sharing excerpts from these interviews, providing you with an inside look at the industry and the unique perspectives of these experts. We hope that these interviews will provide you with a deeper understanding of the industry and its potential, and that you will find them as enlightening and enjoyable as we did. So, join us as we continue to explore the world of Super Apps together.

Janet Young is a member of the senior leadership team under UOB Group CEO, leading delivery channels and digitalization initiatives to achieve UOB Group's vision of a premier regional bank in Asia Pacific. Working with internal business and technology partners, FinTech and ecosystem players externally, she champions collaborative business models to drive better outcomes for the bank and its customers in the digital age.

Janet has more than twenty-five years of banking and treasury experience with Bank of America, HSBC, and Philips Electronics. She serves as a board member of InnoVen, FinLab, Accuron Technologies and Future Economy Council (Modern Services Sub-committee).

Q1. Super Apps are gaining popularity across the globe, especially in Asia and emerging markets. What are your views on them?

- Super Apps refer to mobile applications that offer a user access to a wide range of services, from food delivery to ride-hailing and other on-demand services all within a single platform.
  - (i) It is this ubiquity that draw consumers to use their newly added services and creates the 'stickiness' to capture and retain users on their app daily.
- For users, they have gained popularity due to the **convenience** they provide—the ability to access multiple services through one single app.
- For its creators, the wealth of data generated via the multiple service offerings makes it a 'treasure trove' to drive more personalized services and offers. That said, data security and privacy remain areas of concerns that need to be addressed.

- The rise of its popularity in Asia and emerging markets is largely driven by:
    - (i) The increasing mobile (smartphone) penetration and internet connectivity rates.
        - ○ Based on an Insider Intelligence report in 2022, smartphone adoption among internet users in Southeast Asia is expected to increase from 88 per cent in 2022 to 90.1 per cent in 2026.
    - (ii) A large, young and mobile-first population.
        - ○ more than half the 660 million strong population in SEA is under thirty years old and smartphones + mobile apps have become second nature.
    - (iii) The ease of mobile payment wallets (a typical embedded finance feature in Super Apps) to serve the underbanked and unbanked population, especially during the COVID-19 pandemic, which quickened consumer cashless adoption in the region.
        - ○ According to a 2021 study by payments network FinTech Boku, SEA has been named the fastest growing mobile wallet region in the world, with the # of active mobile wallets expected to grow fourfold to reach 439.7million by 2025.

Q2. How do you see the role of Super Apps evolving in the financial industry and what impact do you think they will have on traditional banking?

- Many Super Apps have started out with a single function that does not involve financial services. However, as they grow in stickiness, it is a natural progression to embed financial offerings such as mobile payments, to loans and insurance, into their platforms.
    - (i) This can increase access to financial services for those who may not have had access to traditional banking

services and provide an additional option for those who already use traditional banking services.

(ii) Closer to home, Super App Grab has also extended its financial service ambitions into offering digital banking services by securing the necessary digital banking licenses in SG, MY, and taken a stake in Indonesia's Bank Fama.

- Apart from helping to keep users within the app, Super App financial service offerings such as payments processing, help to create additional connections between buyers and sellers (merchants) to form a marketplace structure.

(i) Super Apps, with their user-friendly and quick setup interface, help small and medium-sized enterprises (SMEs) overcome key challenges such as payment processing to make it more accessible for merchants to sell online.

- The impact of Super Apps to traditional banking include:

(i) Increased competition and disintermediation of traditional banking services

  ○ By offering a wide range of financial services offered by traditional banks, including digital payments, money transfers and even lending and insurance services, Super Apps are a potential threat by distancing FIs from their customers and increasing competition in the financial services market.

  ○ Generally, the more comprehensive a set of financial offerings a Super App can provide, the higher the threat and risk of competition it poses to the FI.

(ii) The need for regulators to ensure that Super Apps in the financial services space are subject to the same standards and oversight as traditional banks in the provision of their financial products.

Q3. Do you believe that there is a Super App monopoly in Singapore? Especially after Grab's merger with Uber in Southeast Asia.

- In the recent years, we see a trend towards a concentration of services within a few Super App platforms, offering various services such as ride-hailing, food delivery and mobile payments.
  (i) Based on a Dentsu Singapore Super App survey conducted in 2021, Grab, FoodPanda and Deliveroo are the top three platforms in usage.
- However, what is interesting to note is that not one of these three enjoyed absolute user loyalty. In fact, 80 per cent of the respondents were reported to be using at least two Super Apps at any one point in time for reasons such as:
  (i) Cycling through the range of Super Apps to check if one offers a cheaper service than another (58 per cent) at the point of purchase;
  (ii) Doing comparisons to find the best deals and offers available (50 per cent).
- This suggests that at this current point in time, there is no true monopoly, rather users in Singapore are highly motivated to move across Super Apps based on consumer promotions and incentives.

Q4. Why do some Super Apps, such as Gojek and Grab, have high valuations even though they have not yet turned much of a profit?

This was the case in the earlier days in 2019 and 2020. However, the situation is not the same today as we see significant valuation correction happening. This point should also be raised or addressed.

- Super Apps such as Grab and GoTo (Gojek) have seen high valuations due to their potential for long-term growth, market dominance and perceived profitability.
  (i) This is especially the case for SEA because of the large size of its digital economy and the growth of its digitally savvy population.
     o Based on Google, Temasek & Bain's e-Conomy SEA 2022 report, SEA's digital economy remains on course to reach $200B in gross merchandise value (GMV) in 2022 and digital adoption continues to rise.
  (ii) As these platforms continue to acquire and grow in 'stickiness' to retain large numbers of users, it creates a virtuous cycle of network effects—i.e. as more users join the platform, more merchants are attracted to the platform itself which again in turn, leads to more users.
  (iii) This large user base and extensive data amassed become very valuable as they provide monetization opportunities in areas such as targeted advertising, transaction fees, etc.
- That said, given the tough macroeconomic environment today, we see significant valuation correction happening to public-listed companies like Grab and GoTo happening already.
  (i) Since these companies have gone public in Dec 2021 (Grab) and Apr 2022 (GoTo), their share prices have tumbled to a high of 70 per cent even as they continue to report losses over the subsequent quarters.

     Unless these companies can demonstrate that its scale and perceived long-term growth can be backed up by a clear path to monetization and profitability, investors remain wary on whether these companies are able to live up to its expectations as a Super App.

They can no longer adopt a 'grow at all costs' model in a bid to increase market share.

(ii) In addition, even as Super Apps are making in-roads into financial services across SEA having obtained digital bank licenses (Grab in SG and MY) or increasing their ownership in banks (GoTo and Bank Jago, Grab and Bank Fama), they will be faced with stiff competition as incumbent banks in the region are also stepping up to the challenge, i.e.

- o UOB TMRW, an all-in-one app built around customer needs to enable them to bank, to pay and to play; currently available in SG, TH and ID;

- o MY Maybank's Maybank Anytime, Everyone (MAE) mobile banking and e-wallet application to seamlessly integrate online banking with lifestyle needs;

- o KBank in TH making plans to turn its 'Make' mobile banking app into a purely digital bank and expanding its network within a 'super ecosystem' with business partners (seven groups including e-commerce and payment, social media, financial service providers, education, public health, clean energy and tourism) to address customers' lifestyle needs.

- o BCA in ID launching Blu, a digital banking app focused on digitally-savvy customers, which gained more than 1M customers and managed $4.6B in transactions as on 22 November since its launch on 21 July.

Q5. Why are Super Apps more popular in the East than in the West? Are there any cultural, regulatory, or market-related factors that contribute to this difference?

- This is likely due to a combination of cultural, regulatory and market-related factors:
  - (i) Cultural
    - o Mobile-first population: Many countries in the East have a higher per centage of people who primarily use their smartphones to access the internet which makes Super Apps a more convenient solution of choice.
  - (ii) Regulatory
    - o Given past cases of data security lapses by big tech companies, regulators in the developed markets have developed a more comprehensive framework to enforce stricter practices in data sharing and privacy. This puts more pressure on Super App companies and their business models as they seek to ensure compliance with the necessary regulations.
  - (iii) Market-related conditions
    - o In developed markets such as North America and Europe, the digital markets tend to be more consolidated with each vertical already having existing players dominating in that space. This makes it fiercely competitive and harder for a Super App to come in and find scale.
      - For example in the US, Uber is reported to have ~70 per cent share of the ride-hailing market, while Amazon holds a ~40 per cent market share in e-commerce last year.

Q6. Are there any ethical concerns or challenges that arise with the development and use of Super Apps, for example, issues regarding users personal data collection and privacy?
Some of these challenges include:

    (i)   Data Collection and Users Lacking Control
- Vast amounts of data, such as users' personal information, browsing habits, location, are typically collected by Super Apps across the multiple services they offer.
  - Users may not be fully aware of what data is being collected about them and/or have limited control over how it is eventually used.

    (ii) Data Security
- Data breaches due to weak cybersecurity infrastructure continue to remain a significant risk. With so much personal information being collected and stored by Super Apps, there is always the risk of such sensitive information falling into the hands of malicious actors.

    (iii) Data Sharing
- Being a multi service offering platform and a marketplace, Super Apps are likely to share user data with third-party companies which will then be able to use the data for their own purposes, potentially leading to unwanted marketing or other types of exploitation.

Q7. What impact do you think Super Apps will have on traditional financial institutions (FIs)?
- Super Apps which provide a wide range of services including financial services have the potential to disintermediate traditional FIs by offering customers more convenient and integrated services.
  - (i) They could also lead to a shift of consumer behaviour and preference towards 'digital-first' solutions.

- That said, incumbent FIs have the following advantages to help them compete and stay relevant in the face of Super Apps:
  (i) A strong brand;
  (ii) A trusted reputation; and
  (iii) Regulatory experience.
      o In some countries, regulatory requirements for financial services such as the need for certain licenses may limit the ability of Super Apps to fully replace traditional FIs.
- Incumbent FIs could choose to respond in several ways:
  (i) Partner with Super Apps to provide the backend services needed for their embedded finance service offerings via white labelling solutions or providing 'Banking as a Service' (BaaS), or to find ways to work together and create even greater value for both set of customers.

  Developing their own Super Apps to improve customer engagement and retention which in the long run could create opportunities for further growth and additional revenue streams.

Q8. Do you see Super Apps lasting the test of time? Will they still be a large part of our lives in the coming five to ten years? And are they here to stay?

- In this part of the world, Super Apps have become a convenient one-stop solution for the daily needs of many and they are likely to stay so long as they **continue to innovate and provide value to users** (via the ability to resolve customer issues quickly, address data and privacy concerns and continuing to build partner relationships

to enhance their platforms) amidst any changing market conditions, regulatory and technology landscape.
- That said, it is essential for Super Apps in SEA to:
  (i) Address the challenges of privacy and data; and
  (ii) Chart a clear path to monetization and profitability to regain investors' confidence amidst the stiffened competition across their multiple service offerings, especially that of financial services.

Q9. What do you think Super Apps will look like in the future, and how will they change?

(i) Super Apps are likely to continue incorporating new technologies and capabilities to extend their range of services to users on one single platform.

(ii) These areas could include **new verticals** such as healthcare, property, education, etc. as businesses in these industries modernize and look towards adopting digital technologies to tap on new business opportunities.

(iii) With the vast amount of data captured across multiple service offerings and the use of technologies such as Artificial Intelligence or Machine Learning, Super Apps are in a good position to offer up **more personalized services** to cater to an individual's needs and preferences and to improve user experience.

Chris Skinner is known as one of the most influential people in technology. He is an independent commentator on the financial markets and fintech. He is an author of Digital Bank and its sequel ValueWeb.

Q1. How do you think Super Apps are different from the traditional apps that you have on your phone?
Super App is all about integrating everything into one app, and each one looks different in different markets. Which is why it's called a Super App. Different markets will have different views around whether that's allowed under their regulations. There's an ambition by Elon Musk to create Super Apps for the US, Europe and the rest of the world, or you could say, Jeff Bezos and Amazon. The issue is that a Super App integrates commercial, social and financial areas, and most governments in Europe and the US would not allow that integration. So it's very difficult, which is why there isn't really a Super App in the Western world.

Q2. I'm glad you picked that example of China, because in the last two years, the country has picked up on CBDC. They have been pushing everyone to use that. How do you see CBDC playing a role in financial inclusion and taking the pie away from WeChat and Alipay?
Yes, I mean, again, picking up on the China theme, it's interesting that President Xi realized that Jack Ma had too much power and has pulled back on that power. In fact, he's pulled the rug underneath Jack Ma, Alibaba, and Alipay by saying you cannot have so much power in commerce and finance and social media. So Jack Ma has now left Ant group. I think the specific thing there is that the government wants to enforce the use of CBDCs, which they're copying using the lead of the Red Letter days for Chinese New Year, which made Alipay and WeChat pay so successful in the beginning. They say that you can swap

Red Letters using CBDCs. I'm pretty sure this is a roundabout view, but my next book is themed around money being centralized or decentralized. And what has been happening is that the flow of commerce through Super Apps is moving into a non-governmental system. The Chinese government is a great example, it has woken up to that fact, and are now saying, we can't allow this; it has to come back into a central system. So that's why they will enforce CBDCs. And most governments around the world will do the same thing as well.

Q3. Right, so moving away from China and looking at the industry at large, how do you think Super Apps monetize their data? And do you have any comments around how that data is being used within their organizations in terms of using AI (Artificial Intelligence) and ML (Machine Learning)?

It's a basic thought, which is, the more data I have about you and your life and what you do online, the more I have to sell about you and your life online. So basically, the monetization of a Super App is all about you as a product, because the more of your data I have, the more I can sell. And this is where there's an interesting movement, which we will see how plays out in the next ten– twenty years, whether the web becomes the web of us rather than the web of companies. So if we can do that, analyze everything, and get to a stage where I own my data and my identity and share it when I feel like it, I can monetize it. Surely that is far better than giving all my data to a company in a Super App, where they monetize it on their side.

Q4. Do you think that's going to happen any time soon, where people will have power over their own data?

Yes and no. Right now, and it may be because of my age, or something else, I'm not being ageist, but it kind of feels very weird. And I think most people don't want anything that's complicated. So for example, we just had Elon Musk taking over X and everyone saying, 'Oh, we need to move to an alternative'. And we have this new thing. Other things are available to us, but they're too difficult to use. The same with cryptocurrencies, to be honest. People were saying crypto is fantastic, and that we should all move to alternative currencies, but it's only the geeks and the nerds and the people who already understand technology that can use those services. So most people just want simple services. CBDCs, governmental organizations, these kinds of simple structures are the things that they find the easiest to use. In fact, there's a thing called the Milgram experiment, which I don't know if you're familiar with, but you can Google it. Basically, it's that if you're given an order by a supervisor or supervisory authority, such as the government, then people will do what they tell you to do, because we're basically herd animals, and we comply with the rules. So most people will not move to alternative systems, alternative currencies, or to alternative ways of doing things, if the government says don't do it.

Q5. How do you see the potential of Super Apps given the cultural differences, do you see any differences between the East and the West? And its impact on the growth of Super Apps?

Yes, I think, immediately there are two big differences. The first is, because most people that I know in Asia, particularly in the big cities, like Hong Kong, or Beijing, or Shanghai, they don't have a lot of space. And you know, most people live in small apartments. And by way of example, the people in the West find it quite amusing that some people in Asia rent their luggage when they travel, because there's no space in their apartment to store it. That's one big difference. So we have loads of space, I have laptops and big screens. But mobiles in Asia makes far more sense if you're in a small apartment. And that's the way you live, mobiles and tablets have definitely dominated the culture, particularly in China, but also in Hong Kong, and other parts of Asia. And the second big difference is the culture and outlook and the way of thinking in that, I always remember in a presentation from a professor from a Chinese university, talking about how they studied the way in which people in America and Europe looked at a picture of a tiger in a river compared to people in Asia. The people in Europe saw the tiger first; they were tracking their eye movements. People in the US and Europe saw the tiger first, then the river, and then the mountain behind the tiger and then the sky and the forest and the whole landscape. Whereas people in Asia saw the landscape and the sky first, and the tiger last, based on their eye movements. Now, I don't know if that's true, but I think it's true that most Asians live in a culture where it's consensus, it's the community, it's the society, it's the big picture. Then you focus down on the smaller details. Whereas people in the US and Europe start with themselves and start with their details and then eventually start looking at the big picture.

Q6. I love this example, because it really defines the way we are willing to just follow the herd and just do what others are doing. I'll just change the focus and talk about the valuation because we read about it in the news all the time. And it comes out that valuations are great, but these companies are not making money. And by companies, I mean, Super Apps, and FinTech firms in general. What do you think about this? So the valuations of the Super App companies are very interesting right now. I remember Goldman Sachs being very upset that they created a digital bank called Marcus that didn't add anything to their valuation when you saw Monzo, Chime, and New Bank getting huge market capitalization based on their speculation. And it's all about what is it worth today versus what is it worth tomorrow? And that's where the market moves, because the markets are always looking at tomorrow, looking at where we can make the most money tomorrow. So when I look at FinTech or the valuations of the metaverse, companies or anything related to hot technologies, such as Artificial Intelligence and blockchain, I go, well, it's all based on the belief that these companies will succeed tomorrow. And yet, today, quite a few of them are on the precipice of disaster, particularly right now with what's happened in the last couple of years with the pandemic and latterly, what appears to be a global recession because of war in Russia and Ukraine. So I would say the valuations are just speculations. And that's what they always are. That applies equally to cryptocurrency or any other area of the market. Last year, you could have had a million dollars or more of cryptocurrency. This year, it's probably worth $100,000 or less. So, it's all speculation.

Q7. Do you think going forward, mergers and acquisition would be a trend? In Singapore, Grab acquired Uber to become to become a key player in Southeast Asia.

I mean, there's definitely a trend to create something that's the one stop portal for citizens. Whether that's run by a commercial company like Grab, Alibaba or whoever, versus the government is one critical question. But for the consumer or the citizen, I think, for us, it's basically, we want to simplify our lives and to stop having to mess around with so many things. So if we can do everything in one app, and it's as simple as it can be, and we just swipe left or right or very near in the future, we just say, 'Hey, do this, do that' based on voice control. That's the way we're going togo. Right now I think a lot of the companies are focusing on speech recognition, for example, within Super Apps. Because if you think about that, then it'll be so easy to run our lives just not having to even text or type, you just talk and meet, it's going to be so much easier. One of our customer obsessions is making customers' lives easy, but governments, unfortunately, are very bad at this because it should be about making citizens' lives easy. And they've delegated their authority to companies like Ant and Tencent. And now they're trying to get it back. And so that's going to be a really interesting conflict in the next decade.

Q8. Yes, you're right. And then there's a conversation about the four pillars of the growth of any kind of industry, right infrastructure, venture capital, ecosystem growth, and talent. You spoke about East versus West. The West is flourishing in the four pillars yet we see no Super Apps. Do you think West is missing out on the Super App bandwagon?
Yeah, and I should also clarify just between you and I, that I hate talking about the East and the West, because it sounds like we're in different parts of the world. You know, I mean, if the centre of my world is Singapore, am I in the East or the West, or in the centre of the world? So it kind of creates this view of a world where we all sit down with a position that says we're dominating or not dominating. The reason I picked me up on this is that in the Middle East, as we would call it, the states of Saudi Kuwait or United Arab Emirates, now want to call themselves West Asia, rather than Middle East. So I feel very nervous when we talk about East vs. West.

Having said that, what I would say is America and Europe, are they losing out? Because they don't have Super Apps? Not really. At the end of the day, the dominating factor for me is Samsung, Google, Apple, and smartphones. Do I really need a Super App? Or do I just need a smartphone, that can link me to all the services I need in a very simple way with touch or talk. And that's where I'm at. I don't think it really is a Super App, what you really need is apps that can respond very quickly to a message from the customer saying, I need this, I need that. And that's what we're going to see in the next few years, particularly in Europe and the US. It already has that to a large extent. But as I go back to saying it's run by corporations, rather than by governments, which makes it difficult.

Q9. Chris, what is your view on emerging markets and frontier markets? Do you see any future trends, perhaps, regulation being more favourable where regulators are playing catch up?

Well, there's a number of layers in there, which is whether we ever get to the stage where there's a Super App, or ability to have an integrated service, which is effectively a Super App, but run by multiple players. That's a global platform. So where we are right now is Amazon, (I'll pick on them as a good example) dominating the American market, the European market, and the Indian market. They don't obviously have much presence in Asia, because Alibaba dominates Asia. But what would happen if Alibaba and Amazon created a global alliance, and then integrated into that global alliance, PayPal and maybe started to generate other services? I've thought about this for quite a long time, which is if you have the big tech companies, developing global platforms and services, and then working together, aligning with each other saying, 'Let's create a global network that's integrated between social, commercial, and financial.' You will have an amazing world and you will have a very difficult world to regulate as a national government, because national governments can only control domestic policies. And now you have global technology players, overriding your policies. How would you deal with that?

Q10. That's food for thought. I think if that happens, it's going to be an amazing time for customers. You don't really

need to worry where you're sitting in the world, because you will have access to everything.

Do you think we will still be talking about Super Apps five–ten years from now, or it's just a trend which will go away?

I don't think we're talking about Super Apps. And I don't think we're talking about FinTech, I think we'll be talking about a world where technology is embedded into our lives, and it's completely invisible. It may not be in ten years, maybe twenty, or thirty years. But I've grown up in a world where we were amazed that we could put a man on the moon and yet now we can talk about putting a woman on the moon, a person of colour on the moon, and a colony on Mars. So the world does actually move very, very quickly. And a lot of the technologies that we think are fantastical today, tomorrow will just be embedded into our lives. And so specifically from my side, I reckon that in ten or twenty years, FinTech and banking will have integrated into something that's part of Super Apps that are driven by artificial intelligence, machine learning, and speech recognition, and are just around us every day as part of our lives. And we won't even think about it. It'll just be magical. And that just goes back to that wonderful quote from Isaac Asimov, I think it was, 'Any sufficiently advanced technology is indistinguishable from magic'.

As the co-founder of redBus, Phanindra Sama is recognized by millions as the man who transformed bus travel in India.

Q1. Phani, given your entrepreneurial background and the way you disrupted the bus booking system, what kind of innovation do you see coming from the Super App model? Are Super Apps really innovative, or is it just a fad?

Yeah, so basically, from the innovation perspective, there's probably not much. First of all, basically the innovation is within the service, right? I mean, a Super App is a collection of services. So putting together a Super App may not be that innovative, but innovating on each of those services is where the actual innovation lies. That's where innovation is. For example, if you take PhonePe, all the things they put on it, and the innovation is not in all the things coming together in a single app. Whereas I think a much better innovation is in each one of those things like insurance products, payment products, and things like that. Yeah, but in general, like, the mall approach, I mean, other than China, the mall approach has not really worked out. And not just in the apps, but even traditionally, if the mall approach worked, the physical model, like where you put all the brands and then it becomes a crowd puller, has worked in the physical offline space, but it has not worked in the online space traditionally ever. I mean, if it had worked, then we would probably not have so many internet companies. I mean, then a single company would have done everything. For example, like when we launched redBus, MakeMyTrip was already a big name. They were advertising on TV, they had transacting customers, it was a big name, MakeMyTrip, Cleartrip, and Yatra. These were very big names when we started bus ticketing. And they had asked us for API. An application programming interface (or API) is a way for two or more computer programs to communicate with each other access to all our inventory so that they can start selling on their website. So it was a big question for us. Because we were almost nobody, we didn't have names, we didn't have as many transacted customers, we were not known there was no brand, there was no trust. So internally, we were worried, right? Should we give our API to them, should we give them access to our inventory or not? But we decided to give it. If we give inventory we may just remain like a back end company that supplies inventory, but the front end may be taken by MakeMyTrip or Cleartrip or Yatra which means the consumer brand for bus tickets will be MakeMyTrip, Cleartrip and Yatra and not redBus, we were worried that it won't be redBus.

And as you know, most revenue money, the value add is in serving the customer and not necessarily in the supply chain. Brands end up making more money than the suppliers; suppliers are replaceable. So we worried about that. But somehow we decided to open up our inventory to all of them. But they never sold more than like 5–10 per cent of what we sell on a daily basis. They never crossed that in spite of their brand, distribution access, technology, whatever it is right? I mean, they just could not cross more than 5–10 per cent of what we would sell every day. Much later in the app space. Paytm also wanted access to purchase inventory, because they wanted to sell bus tickets when they were creating the Super App on Paytm. They want to sell bus tickets on Paytm. Again, we gave access to Paytm in spite of their aggressive offers and deals. They could not really reach significant market share in bus tickets, and much later they reached. I mean, today, they may have some good presence in bus tickets, but I don't think they are, like, more than 30 per cent of what redBus is; like they're three times smaller than redBus, in spite of their very aggressive offers. I mean, they may be much smaller if we look at the latest data, if they dropped their offers. So all that shows that in the internet space, maybe Super Apps, that the mall approach is probably not really working out.

Q2. Do you think Super Apps have managed to disrupt the e-commerce, bringing more traffic on their app?

See, unlike in offline retail, in online retail, people want to go to experts. In the offline retail a mall works because of accessibility, like you may have a bigger Nike shoe shop, like 10 kilometers away, but if there is a smaller Nike shop, like 2 kilometers away from your house, you're more likely to just go to the 2 kilometers away one right? I mean, you may not want to drive all that far. But on the internet, right, the Nike shop in a mall website, there is no distance. They are as easy to access, like the Nike website is as easy to access, compared to another website. So why would I go to a mall? Right? I would directly go to Nike. Like if Swiggy is available on PhonePe, I would still directly go to Swiggy. Why would I go to PhonePe and then go to Swiggy? I think even PhonePe sells redBus, but the transactions are insignificant.

Q3. Phani, you are based in Bengaluru, which is the startup capital of India and as a Venture Capitalist, how do you decide where would your money go?

Yes, so it depends on what kind of unique offerings that they have, like, something that they created by themselves, differentiated products. Are they selling somebody else's products? Like, if a Super App has BigBasket, Swiggy, redBus, MakeMyTrip, and something else. If they're selling some other brands, then it may not be worth it. But if the Super App B has its own products, which are differentiated and they have expertise in that, maybe that's much more likely to be successful. They need to bring their expertise. It's not just reselling something else. And the other thing is, if a Super App is aggregating smaller, unorganized players, then again, it's good, right? I mean because it brings in trust, credibility and all, like we've seen in Amazon, right? I mean, why would I just go and buy directly from a maker of let's say, a table? Why would I buy it on Amazon? Because there is credibility and trust that it will be delivered on time. If I want to return, I can return. Even though, that seller has its own website, I'll still buy through an aggregator because of those things. But when there is trust already there with a brand, I mean, it's less likely for me to go to an aggregator or a Super App.

Q4. Any views on Indian Super App market?

I think more than the data, one of the characteristics that is needed for the Super App is high frequency of usage. Multiple times within a day. See PhonePe and Paytm; they are way ahead because they are used multiple times in a day. In China, TenCent is that Super App because it is used multiple times in a day. So that way, I think WhatsApp has a much higher chance than Tata CLiQ, Jio, etc. I mean, see, I think there is a very different commerce that happens, right? I mean, if you look at Google advertising, and Facebook advertising or Instagram advertising, right, I mean, Google is very purposeful. Like when I'm looking for something, if the advertisement comes, I may go and end up buying it. Whereas on Instagram, and Facebook, I'm not looking for anything. I'm just on the platform, and it's more of an intuitive, impulsive, buy. I just look at something, and I feel like, 'Oh, I wanted to buy that table mat'. I see a beautiful table mat, okay, let me just go and buy it. There is no intention, but people go and buy. So I think in Super Apps also, maybe that bifurcation will happen. The kind of commerce that is possible on a Super App versus on a regular app a day will be very different. The types of things that you can sell on Google advertisements are different from what you can sell on Facebook and Instagram. The approaches are different. And both are massive, both are big. Both are making a lot of money, lots of money from selling products, both the Google platform and Facebook platform, but they are very different. Underpinning is very different for them. So on Super Apps it could mostly be I think, intuitive things, more impulsive buys could be more successful, because they're not looking at it. For example, if you look at the group buy, which was launched on top of WeChat. I mean, that's not purposeful. The content-based selling, the small mini apps that Tencent has, right, I mean, that's again, impulsive; you read an article or something about lipstick, and you go and buy lipstick, you never thought you wanted that lipstick, right? So it's very different.

Q5. Okay, that's an interesting point of view. India is witnessing huge growth story and has launched E-Rupee: Central Bank Digital Currency (CBDC). Do you think CBDC would challenge Super Apps or the payment apps of that nature, especially in the payment space?

Not necessarily. But they are ripe for disruption. I mean, when Unified Payment Interface (UPI) came in, PhonePe came in aggressively and kind of took away market share from Paytm. Now Paytm, PhonePe, and Google Pay are all very conscious about it. So they may not leave the opportunity for somebody else. But there is always a very good opportunity for some new guy to come and take that market. But investors are also very wary now, because the payment apps are not making money, right? So you know, they are very wary of funding such business models, as a result I'm not really sure. I mean, if it was before the pandemic, then there would be money that will go into apps, which will build a payment mechanism on top of this new system. But today in the given macro situation, it is unlikely but, you never know. You never know if the existing guys don't innovate then there's opportunity for new guys. Paytm was gung-ho on the wallets when UPI came in, and PhonePe came in and disrupted that.

Q6. Digital Financial Services (DFS) is a means to serve the underbanked and unbanked population. What's your view on the nexus between Super Apps and financial inclusion?

Okay, so I think if we talk about the payment apps as Super Apps, or in general, like you refer to the data as oil kind of a thing. So the problem with the unbanked is that there is no data for them. And all the financing activity works on data because of the scale at which it also has to work, right? When you go to the masses, you have to have automated ways to evaluate creditworthiness and things like that. Which means it has to be based on data and algorithms. Given the size of the funding or finances that are given out also that's another reason why it cannot be manual based, it has to be based on some data. And that way I think Super Apps are useful because they have a lot of data, micro transaction data or even otherwise. What food they are ordering on Swiggy (an Indian online food order and delivery platform), there are algorithms that can correlate their willingness to pay back the debt and things like that. And that way, maybe there is a silver lining. Maybe it's beneficial.

Q7. Okay. And do you have a favourite? Is there one that you go back to from time to time? Multiple clicks in a day?
Yeah, it's WhatsApp. It's Google Pay. But Google Pay the challenge with all these things is that if Google Pay supposedly had some service that is not there elsewhere, then, in my mind, Google Pay would be that Super App. But right now Google Pay has I think some bill pay, some DTH (Direct-to-Home television) recharge, and things. I never look at Google Pay for that. My mind has already mapped some of the service for that. There is Paytm, I use Paytm, for my toll, the Fast Tag, I use Paytm, for payment for the police tickets for the car traffic chalans, and similar such events. So before Paytm had it, I didn't know how to pay for it. I think instead of selling bus tickets, train tickets, which are already elsewhere, if they start offering these new services, then it will become a Super App in our mind.

Q8. Okay, so you don't see Paytm like a Super App right now.
But I'm using it for more than payment, right? I mean, that's probably the Super App in my mind. Because I use it for Fast Tag, I use it for traffic chalans.

Q9. Indeed, Super Apps simplify the user experience, and put the missing pieces of the puzzle for the users. What are your views on the business model and market valuation of Super Apps?
I mean, I have no comments on it. But I think Paytm has just started taking Merchant Discount Rates (MDRs), a fee levied on merchants for the payment processing services through debit and credit card transaction) for transactions. So if that's the case, if that is mandated by the regulator, then I think Paytm's valuation will very well be justified. In fact, I started hearing that in the market that since Paytm started taking MDR, people are thinking whether it is valued cheap now. People are wondering if they should buy the stock? Because of the MDR, I mean, see at mass scale, right? At a large scale a small change can mean a lot. I mean, we've seen it. On Facebook for example, everybody was wondering, how would they ever make money? I mean, in private valuations, is the company highly valued? Until then the whole world was used to the user coming and searching something, expressing their intent and then an ad being shown. Whereas when Facebook scrolling, I mean, you're not expressing any intent. You're not saying I'm looking for this. I'm looking for that. No. So people questioned whether Facebook would be able to make any ad revenue ever, right? And then it just switched on the ads, and started making billions. So on a large scale, right, I mean, one small change can make a big difference. And it's happening as we speak, in Paytm's case once the MDR has come in.

Q10. How do you see the future of Super Apps in light of X's recent moves, do you think they could be the next Super App?

Not necessarily. I mean, whatever X is doing its doing within its scope, right? Super Apps are very different. Like, if X introduces NFT's and super fans, or whatever followers and things like that, it's within the scope, it's a feature. It's not like, it's not making it a Super App. If X starts selling like Tesla cars, then we wonder if it is doing something beyond their remit?

Q11. So, in the next five to ten years, do you think we'll still be having this conversation about Super Apps? Or do you think we would have moved on to something else?

I think we will probably move on.

Q12. Any emerging technology becoming mainstream in the future?

Yeah, there's a good chance for metaverse. Blockchain leading to the metaverse, right. I mean there's certainly a very good chance of crazy innovations happening in blockchain. Of course, tons of applications built on top of GPT, Open AI. And the world is changing so rapidly in front of us right now. Like, five years down the lane, something that doesn't exist today will become very mainstream, very common, like you would use it and not even think that it didn't exist five years ago. A great example of this is Zoom. Today we don't even think of Zoom. Probably five years ago, none of us were using Zoom or UPI; the world is changing so, so rapidly. Maybe everything that is composed will be done using GPT and Open AI. You won't even recognize the transition and it could even become a mainstream way of writing stuff, or communicating. It is just changing so rapidly.

Patrick Pun is currently a tenured Associate Professor and the Programme Director of Master of Science in Financial Technology (MSc in FinTech) at School of Physical and Mathematics Sciences, Nanyang Technological University, Singapore.

Q1. What is your understanding of Super Apps? How do you think it's different from a traditional app?

As an app user, I would say a Super App should mean something that can, facilitate our lifestyle and you know, allow us to use one app to do many things in life, like finance, government service, hospital service—all are in one app and we can even avoid multiple verification. So, perhaps just like the SingPass in Singapore, we can just log in once and within a certain period of using the app we can actually just use different services without any obstacle. I mean, we don't really have to be signed in and we don't even need to bring our physical ID card and so on. I would think that a Super App should not just be used to complete a service, it should also recommend me, the user to do something that is beyond my planning. However, that can eventually improve my current plan. So something like this? For example, a Super App could suggest a restaurant during lunch time that may be suitable for me and if I go to a restaurant, what type of dishes would be good for me? Something like this will be from the lifestyle perspective, but from the financial perspective, I would think that. A Super App should be able to allow us to easily access financial services because I think in traditional banking, traditional finance industry. So there are many services that are quite complicated, and if we have an app that can actually make all the things easy to access, easier to use than that I will think, that is what I would call as a Super App, so we don't need to have a consultant or have a professional to explain to me what is this? What is this? But instead apps can help me to simplify a lot of things. However, with the technology we can also make sure that the process is rigorous and complies with the regulation policies then that will be great. So that's why I think I cannot give a very precise definition of the Super App but here is how I understand how a Super App is.

Q2. I think you unpacked it well, Professor, you are dialling in from China and perhaps using WeChat or Alipay back home. Did you see it as a common trend among all generations, or is it just the millennials or GenZ who are catching up on the trend of Super Apps?

Yes, I think so. So I can totally feel that, especially in China and in Macau now where I reside. Basically we will use Alipay or some other e-payment apps to complete all the transactions. That means we don't even need to bring any cash, right? And then we can just complete most of the transactions and so on. I think Super Apps should also be able to allow us to, I think one key principle is that when I employ a few services, let's say I go to a hospital and then I need to pay my bill. Perhaps the reservation system and the payment system are integrated so perhaps there should be something like this and finance will somehow emerge in all the aspects of our life and that's why I think Alipay or any e-payment methods will be like daily tools for us to use in a Super App, something like a gas of a car.

Q3. So Professor, you spoke about the ease that it brings for the consumers. And on that note, I wanted to understand your view on their business model, how do these apps make money? Do they monetize on the consumer data?

I think this is quite a difficult question. In the sense that the business model is not that straightforward and it's not like the traditional industry, we have supply demand and we can easily calculate the revenues and cost and come up with the profit and loss and so on. So especially when a company tries to develop a

Super App, I think the market share is a key and after that, this is not my expertise, but I think market share is key and then once they can take the major proportion of the market share they can charge from the commercials, charge from the retail stores, and even can charge the government for providing such services. So I think Super App will have a lot of ways to do this, especially if you occupy a large proportion of the market share, then you will be able to negotiate with the stakeholders on What can? Can they get better? But I think that it's too commercial or I don't know. But you can easily think of one way to make money by the Super App companies that they can advertise right they can put advertisements on anything right. So that's why I think that will be the major business model, how they make money.

Q4. And in terms of what you just mentioned, the market share, so I believe market share is the key because that's how they're able to diversify their portfolio, bring in more services, but it's also a chicken-and-the-egg situation, right when they're starting out, they might want to promise the universe, but they don't really have the market share. So how do you think the journey takes place for Super Apps from day one in terms of bringing services, but also trying to get a bigger pie of the meal?

In my opinion, that is the part that the company needs to reserve a significant amount of capital to advertise their app or they may want to acquire some mature companies that already have a significant proportion of the market share. So for example, if there is some app that has been quite popular. As a very extreme case, if now let's say Facebook wants to develop a Super App or Google wants to develop a Super App, then it is much easier. But if we just talk about the original company. So we may also have some app that has been quite popular locally, but they do not have those kinds of other features supposed to be in the Super App, right? Just like what we described. So I think one way to do it for those companies is to really acquire them, right? If you have that money right? Then you can acquire them. If not, if we really start a company from scratch, then I think the company will suffer quite a long period of, you know, starting up they need to have many promotions. Like I think I refer to the experience that I had with Fave Pay. Fave Pay is doing quite well actually, especially at the beginning as a user I find that Fave Pay in Singapore has a lot of discounts or vouchers that are provided to the users. So just like Uber and Grab at the very beginning they are going to give us many promotional codes and so on. So, I believe all these companies have one common feature: they want to first occupy a significant proportion of the market share, right? So that's why somehow that is the cost that they need to pay. They clearly understand that during that period of providing such large discounts, vouchers to their customers, they are losing money. But their aim is to first take care of the market share and then they can develop what they want to develop, right? So that's why I think even for Grab, many people are using it but they are losing money, right? They are losing money, especially in the very first few years. They are losing money, losing a lot of money. But that is ok because they will be valued high because they have a large group of customers.

Q5. And you nicely touched upon the big giant companies like Facebook and WhatsApp, and we cannot really talk about any Super App without looking at how they're going to impact the tech industry. So how do you think Super Apps are going to be operating in the ever-changing world going forward?
I think before a Super App is formed it should be a small app, a normal app, but that should be a very innovative app that can really help people to improve their life. And then from there they

can develop it as a Super App. So that is one key bill that I have and that's why I would say, um, so I will not be able to say what part that the model is lacking of why I have some ideas but I think that is not my expertise just from my perspective of being a user then I believe that app that is going to be popular should be innovative because since app development got popular, there are many apps already. So if you want to develop a good app, you have to be very innovative and everyone just knows that they need that app. So I mean it's a bit abstract, but that is how I describe a good app. And then from there you have a solid foundation of the customer base and from there you can develop a Super App.

Q6. In Singapore, Grab offers food, transport, retail, and payment processing. How do you think they are able to manage so many services?

So of course, they have many services, but they will have many different departments to take care of each of the services. I think the more difficult one is how to integrate them so we can have a GrabPay department. We have a Grab Food department but how can we integrate all of them? Grab should have done some very good job to make sure that someone is standing at a very high level to oversee all of them. So how to integrate different services will then be a key, in my opinion, is a key job in developing a Super App. Now I'm not sure about the details, but I believe that Grab has been doing quite well about the integration.

Q7. What do you see as the key to success in a Super App?

First of all, there should be consistent improvement in the services, because the Super App should be focused on certain services, however they want to incorporate more services in order to be a Super App. And second, I think there is a very diverse portfolio in financial management. So now as a Super App or as a company developing a Super App, I am going to touch on many services, but, some of these may lose money, whereas some may make money. So I think a company that wants to develop a Super App should not just focus on those who are making money. And that will basically be like how in finance, you are putting all the eggs in one basket, then that is no longer in the spirit of the Super App. I think Super Apps should be diverse. So, you also don't know which one will then be useful, right? I mean will then be appreciated by the customers. Which service will be the main part that keeps their customer. So, we don't know. That's why you are going to develop all different services and then I think the key again is to integrate all of them into one app to make it easy to use. And I think talking about all this, right, my whole idea or whole thought is still about how to maintain the market share, right, as a Super App, you want data, right? So, you want to have all the customers' usage on different services and from that data there will be a good loop for the company to recommend other services to the customer. So, I think there is a good loop where you are going to build this loop such that you will have customers and customer profile data with the good data you can attract more customers and so on. So, we are going to build or we are going to establish such a loop. If you didn't do well in certain steps, I believe that perhaps the app is still ok for certain services, but the risk is still quite high in my opinion, yes.

Q8. Yes, totally agree with your views on diversification. We spoke about data, how do Super Apps leverage data and use analytics to drive user engagement and retention?
Yes, I think that is the part that academia is also working on. So now we will have more types of data and it's cheaper and easier to store the data. How do we extract the important information from the data? So for the use case in companies, right, I think just like in e-commerce, we can see a lot of recommendations right? That is part of the theory called the Commander System.

That is quite doable and not difficult to make use of the data, but in order to build an accurate recommendation system the data, a massive amount of the data is a key, so we really need to know how we can find someone similar to you right? So that is very crucial. If the data set is too small, then the recommendation is not accurate. So that's why we need to have data. I think the key is about how we can do so because as an individual, you will not be able to know how to make the optimal decision. But with data with some others experience, the app may be able to advise you to do so, right? So that is the key and how we make use of data. And I think the data is more of an assistant. You may not really need to fully follow the suggestion, right? So it's always a recommendation only. That is something we can make use of from the data. And I believe that we can also learn from the data and its analytics. So far I think I have just come up with this recommendation system. It's more from the retail perspective, but with more data, we might also need governance. Imagine if the Super Apps are regulated by government. So if you have data, you basically know the trend of the people, what they are thinking about or what they (we don't need to know the individual one but we can know the whole group or large crowd of people) are doing and so on. So that will be something we can learn from. We can use privacy preserving technology to preserve the privacy of the customers, then that is also doable with the new technologies. So yeah, there will be a lot of potential. Just like at the beginning of answering this question, academia is also looking into it and seeing how we can assist the companies to do more. I think it's always about innovation so we have the data, we somehow have technologies, but somehow the missing part is the innovation. But the innovation part could come from academia or could come from anyone, any user. Yes, that's why I think the idea should, I mean the data should be collected first and then we can always try to build a model for it. So there is the usage of the

Super App. If you have one app that can integrate all the data, have all kinds of data in one platform then we will have great potential to develop something useful, right?

Q9. Easter world is dominating the Super App space as date privacy is not a key concern. Western world is guarded when it comes to data sharing. Do you reckon this could be a reason for lack of Super Apps in the western world?
Yes, I think this is a very good question. You make a very good observation and that's true. So basically you know, just like WeChat was developed in China and also there are some other apps in Singapore that is, I mean, all of them are mainly in Eastern countries. But for Western countries, I think the main concern is all about their privacy,

For example, in Singapore and China, we need to scan the QR code to record where we went during the pandemic. And that has a very good reason to do this. But people in the west, may not choose to do this. So that somehow violates their core values and I totally understand. That is why I would like to point out that the new technologies are important. New technology allows us to have security of the data on one hand, but on the other hand, we can also analyze the data.

So that seems to be quite impossible, but that is what I mentioned, the privacy preserving technologies. We can do machine learning, we can do deep learning without knowing any individual data, but the data is first encrypted and sent to the cloud. That's why each of the individual data is not accessible password protected. We can analyze it on the cloud, so that will be very helpful. Especially, I think now we are talking about how we can build our Super App so that maybe each subscriber of a company is willing to share their data to this Super App, right? So there is the key. Perhaps we can just think about how we can allow it. I think the Singapore government has already adopted this technology because for financial data we have the SGFinDex (Singapore Financial Data Exchange). This enables integration of the financial data from different banks for each person. Previously this was a very large hurdle in the sense that different companies, different banks—even within the same bank, the data from different departments within the same bank will not communicate. So they will keep their data by their departments because its transition will somehow incur some leaking and so on. So this is about privacy problems. But now we can already use some technologies to allow us to share the data for different banks. So I think we are going to adopt similar technologies in Super Apps such that we can assure the user that their information and data will not be leaked to others while we can still make use of the data to analyze.

Q10. Which is Your Go-to Super App and why?

I have used WeChat. As I said, I'm not so clear about the definition of a Super App, but I think WeChat should be considered as a Super App. I also think that a number of apps in Singapore may also be considered as a Super App in the sense that they provide certain convenience in using certain services. For example, SingPass is linked to many government apps, but there are many different apps so I'm not sure whether I should call it a Super App or Super Apps, right? So anyway, I think the idea is that I can verify my ID using hospital service with SingPass. I can pay for my test with SingPass without going to the authorities. I think it is very convenient in Singapore. I appreciate the convenience and the good thing is that it is quite easy to use. For WeChat, there are many apps in WeChat, so called 'services'. You will be able to open a menu and from that menu all the sub services are categorized and then you select the corresponding one. Perhaps you want to pay your utility bill. You want to top up your EzLink (rechargeable contactless smart card for public bus and MRT in Singapore) card and so on. All these are within one app and of course they also have the WeChat Pay, right. Basically, when you use all these services, you can just pay within the app. So you don't need to leave the app. And I think it's also very secure. You know, 1 billion people in China have been using WeChat and I haven't heard many scandals about using this app so I think that is quite secure. So in WeChat the experience was quite good, but, there was also a bad experience about it being too powerful. Too powerful in the sense that if you, I mean the app somehow could be under maintenance, right? When you do all your services in life in one app, then again there is another way to put all the eggs into one basket. If something is wrong with the app or your phone is somehow not compatible with the app, like when you click on the app then it will jump to the home page. It it's too powerful it will also bring some inconvenience in some cases. So that's why I think the server maintenance is very important for a Super App because you want people to come on your app and use your app to do all the things. But yes, perhaps you will be able to attract customers because you have many services and each of the services you do well, you can attract corresponding groups of the customers. They will just use your app. But you will also receive a lot of complaints because each of the services will receive a significant amount of the complaints. But in my opinion, my experience was quite good, yeah.

Q11. Your views on the future of Super Apps?
It's my hope on WeChat that they can have more accurate and positive information. I mean so one of the ways that people use WeChat is to use it like X, right? People will learn the information from WeChat and so on. So that's why I would think that if that has been a platform that many people are using,

I hope that it will be a platform that allows people to learn the correct information. And I believe that will also be good for the company, so.

It has to be positive, in my opinion. I'm not sure whether I think my discussion here is really very personal. I mean not speaking from a business man perspective, perhaps from the business person's perspective, I would say yes, the app may incorporate more services, for example, I think, what they can do is to educate the elderly. For instance, in China they can really teach and do more volunteer service to teach the elderly people how to use it right? I think the young people know how to use it but there are still a large group of people who don't know how to use those Super Apps. So I think that is what they can do. But since now when I view this Super App, I will be a bit worried about what kind of information the kids will receive. Because Super Apps are so powerful, I think sometimes as a company they need to take one step to ensure safety. I mean they need to go an extra mile to prevent the app from being used improperly. So that is something I would hope the company can do.

Especially, since I am in the education industry, I don't want my students to learn something fake from the Super App.

Q12. Yes, I definitely think education for the elderly and for all the young kids is very important for the sustainability of the Super App. Just looking at the big picture, do you think five years down the line we will still be having a conversation about Super Apps or is it just a fad that will die down? Is it just like a new wave of cryptos or something in Fintech, do you think it's going to survive the test of time? And will all of us still be using Super Apps?

I am not an expert in this field, but in my opinion, I think each country may want to have one Super App. It does to another extreme when compared to the crypto currencies or crypto assets. The Super App is more super centralized and it will be good for governance and that's why that will also provide convenience to the users. So in my opinion I see a lot of benefits to having a Super App. But if we consider all those concerns that we just discussed, like for the Western people, they don't want to share their data with a single company, right? So a Super App cannot really survive. Depending on the technology, the one thing we can do is convince people that the Super App is also super secure, safe, and that their data will be preserved very well. Currently, the Super App has this concern. If we eliminate this concern, it seems that, in my opinion, it can survive. From the business model, they will always be able to find ways to be sustainable so they can just run the company like Facebook and Google, they will be able to survive from the business perspective.

Q13. Yes, you're right. Professor, the way you drew the parallel between the centralized approach of a Super App versus the decentralized approach of a cryptocurrency. I'm thinking particularly in the context of China. We have CBDC (Central Bank Digital Currency), the digital yuan and the Chinese Government giving away the red packets. To prevent dollarization and to take things in their own hands, right? But at the same time I think it's also to take things in their own hands, right?

To be the honeypot of the data and not let private players like WeChat and Alipay dominate the market, do you think CBDC is going to be a competition to Super Apps in the future?
I understand what you mean, but that is still not comparable at the moment in my opinion. CBDC has been launched and is used within a large group of people. Right now the Super App makes our life easier, but if we use another financial system, then I think

at the beginning it will be more complicated and I even think they are not competitors. They may be able to, you know, incorporate them so they can be integrated into one Super App. So the idea of the Super App is about how we can integrate all different services into one place. I mean, it's a kind of a new initiative, right? That's why I think now that they are not still not comparable. At least I think I would want to see whether the centralized currency is really used, then perhaps I'll have more of a perspective on this.

Serey Chea is the Governor of the National Bank of Cambodia.

Q1. Which Super Apps do you use the most on your phone? Apps like Grab, GoTo, AliPay, and WeChat...
Well, I have Grab, I have WeChat, which I rarely use, and I also have Taobao, which I also rarely use just to see if there's a cheaper version of something that I want because it's so easy to just take a snap of the object I am looking for. Overall, I think that they have a very good search engine. There is also Uber.

Q2. What do you think of Grab? Grab has transformed itself from just a ride-hailing app to food, to payment, and now they are also doing robot deliveries. Do you use it for all these services or just a few of them?
Only a few services. I don't think anyone can use everything.

Q3. Moving to the Cambodian market, do you see a lot of Super Apps in the market? Do you think their growth is due to Bakong e-wallet, National Bank of Cambodia (NBC) project?
Bakong is only for payment; the other services are not payment-related. So it's not something that our regulation has impeached the development of Super Apps.

We do have Nham 24 where we can order food delivery, and we can order groceries for delivery. They have a couple of other services as well, they do transport, they can pick up stuff from your house and deliver it to your parent's house, things like that. They are doing quite well.

Q4. Cambodia has one of the highest internet penetrations in the world. Mobile connectivity is great, combining it with policy, regulation, and infrastructure, all the support pillars are existing in the market. The penetration of Super Apps,

for instance, Grab, GoTo, AliPay, and WeChat, doesn't go well with the growth story of Cambodian Fintech. What do you think is lacking there?

Well, big fish can live in the ocean, but they have a hard time living in a small aquarium.

So, I would think of Grab and Amazon as big fish who can maneuver big markets. But in markets like Cambodia, it will be a bit more difficult for them. They will need to understand the local context very well. This is probably the reason why they can't pick up here. But I mean, the local Super Apps are doing quite well. In terms of food delivery, Foodpanda is here but they are not doing as well as our local Super Apps, Grab is here as well, but they are not doing as well as our own PassApp, which is the tricycle rickshaw transport app which is doing super well. I mean I haven't seen a financial statement or anything, but this is just from interactions I have on the street with people who say, 'Oh, I will call the PassApp taxi, and I'll come and see you.' So, from these kinds of interactions, I conclude or assume that they are doing much better than Grab.

Q5. When you say Grab is able to survive in the big ocean, are you referring to markets like Singapore, Indonesia, and the likes?

Yes, probably.

Q6. Any particular reason why there's a disparity in Super App growth between Asia and Europe, why are these two parts of the world so far wide apart?

Personally, I think that from adversity comes creativity. I suppose that from the Eastern part of the world, we are facing a lot more challenges, and so we try to be more creative than the Western world. Again, the fact that we don't have any sort of existing providers, I mean, in the West, they have very specialized players so it's very hard to break the system. Whereas in, say, Asia, we don't have that kind of specialized service provider, and so when someone is good at something they keep adding on new things to them, and that's how I think it's that the Super App is growing much bigger in Asia than the West. And also probably because of the regulatory framework, maybe it's the intentional lack of it that has promoted this innovation.

Things sometimes develop much faster than the regulator realizes. In the West, they are always ahead, there's always something that would obstruct certain services from compliance. There are so many rules that can cause start-ups to have difficulties in navigating. Whereas here in Asia, Alibaba and WeChat can develop so well in China. And it comes off due to the lack of regulation and oversight, not intentionally, but because the regulator didn't understand or foresee that they are going to be so big.

Alibaba reached a certain sort of peak where they can use data to come up with consumer risk profiling and can write unsolicited credit to their customers.

That's when the regulators start to be worried, but before that, nobody would have thought that there is even this possibility.

Things like if I buy something from Alibaba and they asked me what time they want it to be delivered and I said 9 p.m., automatically the system will conclude that I must have a job.

So this kind of analytics and machine learning is sometimes relaxed or under oversight from regulators. I think now they have become much more aware of what's happening. So, this is one of the reasons I think why Super Apps were able to flourish in Asia.

Q7. Now that you talk about the nuances, especially the data and privacy, Western world is unwilling to share their data. Asians on the other hand, openly and willingly share data in return for loyalty points or discounts. Do you think this mindset also plays role in the disparity?

Yes, well, that's part of the unintentional lack of regulation. And again, the example that I gave, after a while Alibaba will realize how much I'm spending on a regular basis, they will calculate my monthly income, and they can come up with certain numbers and say, I can give you a credit line up to this much. This is because I do know how much you are earning from the data that I computed in my system. So with all this information that I deliberately, unknowingly gave away, they can do a lot and they can create a lot of services. Now, do I necessarily give this away? Am I necessarily aware of what I was sharing? Not necessarily. Are the regulators concerned about it? They may not even be aware of it until it's already quite late. Data privacy, I would say, is kind of a rich-country problem where you reach a certain status in your life in your culture or your civilization. You become more self-aware of your privacy. But if you have nothing and you can exchange your information for free service or points that you can exchange into goods, or whatsoever, I mean, why not?

Q8. In other words, you're saying that most of the Asian regulators are playing catch up, which is good to help these fintech's thrive?
You basically mean to ask why is it that they mostly come from our part of the world rather than the West. But is it necessarily better? I don't think so. It's debatable. Because as we leave it open for too long, of course, issues arise.

Q9. Do you see that changing any time soon, this kind of dynamic?
People will be more aware of the kind of information that they're willing to share. They will control it well, you know, a small service, but can give them everything now. It's like, okay, if I give you these, what do I get? People are more sort of aware of the importance of their data, and how much they want. So that's definitely possible.

Q10. Do you think Super Apps will venture into the space of remittance soon?

It will eventually come but then it depends on how much regulators are willing to let them do this. What I see is with a Super App, it's going to be very difficult to challenge them as they venture into every space in every industry. So if they go into payments, they will have to fall under the financial regulator. If they are into e-commerce, they will have to fall under the consumer protections, quality control, or commerce custom department. If they're doing delivery, they will have to fall under the transport department. If they're doing so, in the end, we have to know, how do you regulate one entity that is doing everything and how are the regulators going to interact with each other. That is going to be a big challenge going forward because human beings are always people who want to leave others behind so that's what I foresee is going to be a challenge for Super Apps but also for regulators to see to what extent you want to regulate. But at the same time promote this innovation and then again they have to balance whether these innovations are beneficial or not to society, the customers, or the consumers.

Q11. Do you think Super Apps play a role in the lives of unbanked population especially women and youth?

When we talk about financial inclusion, we are talking about access to financial services, insurance, deposits, and all that. This is usually a very, very regulated industry. If you try to sort of mix it up in a Super App platform, and I do want to see the auditing, then you are going to have difficulties regulating what they're doing. Ultimately, it's going to affect the company. If you restrict this, it will impact them and then vice versa. I know, there are some that are doing this right now, but I think we'll have to think carefully about the risk of such services being included in a single Super App.

Q12. When you wear your regulator hat, do you think there were certain things you should have done before the advent of these Super Apps? Do you think something should have been any different from what it is right now? In terms of regulation and policy?

I think, sometimes we need to sit back and watch what's going on. I think we've been in touch with them, and so far, so good, in terms of catching up and intervening on time. So looking back now, maybe, I don't know, from now on, maybe I would see something that I would think I could have done better and intervened earlier. But for now, the fact that we kind of laid back and sat and watched and let it develop has been quite beneficial for the industry.

Q13. Do you think we will be having a conversation on Super Apps five years down the line?

Maybe some of the services would remain, and maybe some would need to be more regulated. When you mentioned that Grab would offer remittance services, I don't know if they already do but if they do then it would be something that regulators should think about how to regulate them. Or if they become, say, an online or digital bank, they don't have a digital bank license and will combine it with what they have at the moment, we have to think about how to make sure that there's a firewall between these businesses. Because, just as an example, if they're a Super App, they have e-commerce, they have logistics, and then they have banking. And then for a regular customer, these are all just one company, right? They don't know who's doing the banking, and they don't understand whatever the firewall between these businesses that the regulator is putting. So, if I buy something from Grab and let's say the product is crap, and the economy changes and the service is bad, then I would be very unhappy with Grab in general, and if I have any deposits with Grab I would probably want to withdraw. And that's the kind of implications that we need to think about. If, say, there is a massive consumer base who's unhappy with certain products and they want to reach out to Grab, who for whatever reason refuses to exchange the products then these hundreds of customers may decide to withdraw their deposits at once. That's the problem.

So there's this sort of interconnectedness that has happened. And it can become quite risky and destabilizing. Say for example, Grab has financial difficulties because their banks cannot provide liquidity; this problem will spill over to other banks who are not even offering all these customer services.

So actually, Grab is doing Robo Advisory (digital platform that provides algorithm-driven, automated financial planning and investment service), at least for the Singapore market. They talk about $1 investments, every time you take a car ride, or you order food, you get an option to park $1 or any amount you want, into this wallet, and they let you make that selection in terms of the risk assessment, and then invest on your behalf. That's currently happening. But you're right, a bad experience can triple down in just leaving the app altogether.

Yeah. So that's what I think is going to be challenging for regulators going forward in terms of dealing with Super Apps and how much super you want them to be super.

Q14. Artificial Intelligence (AI), Machine Learning (ML) and data—where do you see Super Apps going wrong when it comes to using all the analytics? And do you wish something was done differently?

I don't know, Super Apps at the moment seem to be doing very well. Something could happen and we don't know what, and that's the thing with Super Apps. We don't know what is going to happen or what could go wrong. Until they go wrong. And then we say 'Ah! We missed that.' Do I see it? No, we will just watch and see but I mean, we have to be very careful when we use machine learning and AI.

There are certain formulas and algorithms put in place. And there's been some concern about data biases which could very well happen with Super Apps. If let's say Super App would have to give credit, and it will be based on like Alibaba or you know, your credit profiling with that particular app. Then it will be very dangerous because it will only be based on my experience with that one particular Super App and it can spill over to other services that I might get, other service providers as well, because then people tend to trust Super Apps data because they're super, they have super data and have everything. And it may not necessarily be reflective of who I am. That's the danger of being super.

Q15. Do you see any challenges for the Super App industry going forward?

Well, there are going to be a lot of Super Apps. People may not want to download all these Super Apps. If it is super, then there should only be one right? Why should I download ten Super Apps? I want to be honest. If there are so many apps, I don't want to download them on my phone.

They have to compete with each other and that kind of contradicts things in a way because they have to compete to be the best standing. But then again, if you have only one or two standing, you have this monopoly happening, and they can dictate everything. So it's a very interesting debate. But the challenge is going to be competition. For certain countries or regions, I think it's going to be a Super App for the world. And again, like I said, about big fish living in the ocean but not in the small fish aquarium. To be a global Super App, there's going to be a lot of competition. But for regional and domestic Super Apps maybe there is room to grow.

Q16. Right, so this is a very interesting debate, right? How would the future look like?

It is all about having a good balance. I think it's like driving a car, you need to accelerate, and you need to brake, right? If you don't then at the roundabout maybe there could be an accident. So, you need a Super App that is innovative to keep pushing on the accelerator, but also you need the regulators to put in the brake when necessary. So that's how it works. I think it's about having a good balance.

Bakhrom Ibragimov joined Molten in 2022 from EBRD Venture Capital which he founded and managed commitments of €500m over three VC funds, investing in companies such as Docplanner, PandaDoc, PicsArt, Trendyol, among others. Previously Bakhrom was leading investments with Virgin Green Fund as a Founding Principal and TLcom Capital as Investment Director. Bakhrom was also Internet Finance Director in Virgin Media and tech investment banker with Cowen & Co and Credit Suisse.

Q1. What is your view on Super App?
I think on Super Apps, it's an interesting model in certain geographies, certain markets mostly I would say emerging kind of markets, large consumer emerging markets. I guess the idea behind the Super App is as follows, right? Consumers use a maximum of three to four apps on a daily basis. Generally, you would use three to four apps, right? So, if you have one app, and you manage to develop an app which has high engagement, then you can cross sell all the products. Usually that high engagement service is something which consumers use often. Most often you may have seen that it is a payment-based service. That's the kind of thing that users use very often. And on the back of that high engagement of cross-selling one service, cross-selling other products can happen as well, besides payments, it could be some other product which again has high engagement, be it food delivery or even ride hailing. So any kind of high engagement gives you the ability to cross sell other products. That's the typical marketplace dynamic.

There is a kind of network effect type of dynamic where If you develop one strong network for a certain service you can cross sell the products and usually that works well in the growing emerging consumer markets because. a number of those services are not available. So both the verticals are open. So if you develop in one specific segment, you can add other segments for which competition is low in emerging markets. Let's say you had payment, you can add food delivery because there may not be many competitors, so they may not be strong enough in that specific segment. So, kind of leveraging that network effect. It's an interesting model and there have been interesting cases, I think more success cases in emerging markets than in the Western developed markets.

Q2. Is data privacy a hindrance?
I think it might be a hindrance. I'm not sure to what extent this is, because I guess you can also kind of opt into these services easily technologically. So technically I think it should be very easy to work on via product, product features. But it must be some sort of hindrance.

Q3. Wearing the VC hat, how do you make a call, what do you watch out for? How do you look at the business or revenue models?
It is pretty much the same. I mean, whether it's a Super App or any other business, in the VC world, what we're looking for is opportunities to obtain a very large outcome. We see investing as high risk, most of the companies that you invest in will not be able to fulfill their plans. So the ones that actually succeed, you want those to be very large outcomes. So that they pay out not just for the capital investment to them, but also payout for the cases which have not achieved their goals. So, in every case, it's really trying to understand, trying to have a conviction that this can be a very large business and I guess in the case of the Super App, it requires the large consumer market, mostly underpenetrated in a number of verticals. So if you say I'm going to build this kind of Super App, then you're basically saying look, I'm going to start with a certain vertical with a certain tool and then I'm going to add other verticals and that basically means that there is no

competition that a lot of things haven't yet come online or have very strong incumbents. That's the kind of thing that you would be expecting.

Q4. Any insight on how Super Apps capture the market share, say for example, Grab?

I think what they managed to do is win the ridesharing. So, once you win the ridesharing, that's their main kind of addressable market, the ride-sharing audience. And then using that to cross sell other products, right? So, I think in their case it's about winning one vertical, providing superior service, and winning a customer base for the one vertical. And then after that cross selling other products to that segment.

Q5. Data—what are your views?

Data is at the core. That is how we expect the companies to be run. So, we're looking for a data-driven approach. Data can help measure different kinds of impacts, test different hypotheses on the customer acquisition and engagement. So, we do look at data very closely and I think one very important metric is the engagement, that the consumers you acquired actually stay and are satisfied with the product. The analytical framework that is often used in mobile. consumer investments is cohort analysis, which is looking at each cohort and trying to understand how it behaves over time, getting better or worse, as well as when you do changes in certain areas, what the impact is on each of the cohorts. I think that's a very important analysis to make.

Q6. That's a very interesting point of view. But in terms of the monetization, how do startups, or the established players like WeChat and Alipay, monetize their data?

What you can basically do with the data is that it enables you to understand what to sell, and what kinds of segments to enter and how much you can spend on acquisitions. Customer acquisition is a very expensive effort, and by understanding the data, you can estimate what the lifetime of the consumer is. You can estimate the revenue that can be extracted from the consumer as well as decide how much you can spend on customer acquisition.

Q7. If you had to bet your money on Super Apps which one would you pick?

Yeah, it's hard for me to say because I've never covered Southeast Asia. I never looked at Southeast Asian markets, so it's hard for me to say who would be the winner.

On this subject of Super Apps, I have seen it in a couple of markets, but I haven't really seen it outside of Southeast Asia. There are not that many examples of this being rolled out successfully. I mean the one that I can think of is in Kazakhstan, which is Kaspi. Then I guess there have been several attempts and there continue to be several attempts to do something like this in Turkey. So, in Turkey, leading e-commerce companies like Trendyol and Giter are still in the process of developing Super App type functionality.

But I think the judgement is still out there to what extent it works for them. So outside of Southeast Asia that are not that many examples, I guess Uber, you know, is obviously trying to develop both ridesharing and food ordering. I understand that they seem to be doing well, executing well in adding the food delivery business, but as I understand, these are not the same type of magnitude as what you have in China and India.

Q8. So what do you think is the trick?

I guess when Super Apps is addressing the competition. In the early years of mobile phone development, there would be very few apps, and people would be willing to try and use any old random app. Today the competitive situation is very different with over hundred thousand of different apps, consumers use only a few of them. So, I think the idea of cross selling, once you conquer one segment that has high engagement, you can try to kind of cross sell other products. That will probably remain, but at the same time every company is always fighting for survival by new startups, right. You can never say that any of the businesses are here to stay for long and be uncontested by new entrants. And that's actually what we do—we look for the new companies, new businesses that actually turn this route and say, actually there is another way, a better way of doing business.

Q9. Talking about better way of doing business, I think the thought process always drifts towards financial inclusion and these days there's a lot of talk about sustainability and sustainable development goals. Do you see any kind of nexus between Super Apps and financial inclusion?

I think yes, definitely. I guess this technology-driven, data-driven approach enables them to better serve consumers, right. The traditional banks, for example, without the technology and without the right data analytics, they are struggling to kind of, to bank many segments of the consumer market. So, this kind of data-driven, and in some cases, AI-driven approach allows a better kind of ability to serve most of the segments of the consumer market. In terms of inclusion and impact there is that. But at the same time, you know, and often people look at it from different perspectives, but you know, for you to have an impact, it needs to be a successful business, right? There is no benefit of that sort without a successful business.

Q10. I'm just shifting gears. Central Bank Digital Currency (CBDC), how do you see Chinese Super Apps dealing with it?

I think on that subject you need to speak to a VC based in China specifically about what the Chinese Central Bank has done and the effect of that. I think it's very remote what I do on a day-to-day basis right now investing across Europe. So I guess you know the digital currencies overall, it is an enormous market that is suffering at the moment but in the long term it has its place. The one thing is that there needs to be real use cases for cryptocurrencies and blockchain, above the speculation on the price of it.

Q11. Your portfolio companies include companies across different sectors such as DGpays for digital transactions, Docplanner for Healthcare, and Trendyol for fashion. Any chance to incorporate all of this into one Super App?

Well, I think you cannot really. I would struggle with the concept of incorporating very desperate kinds of solutions into one Super App. So, the healthcare with the fashion, I think it needs to be a similar experience. I think mostly it needs to be a very transactional based online experience. Getting healthcare into an efficient transaction based experience, I think it is more and more challenging I would say. I would think it has to be kind of a similar consumer experience along the lines of those verticals.

Q12. As a venture capitalist, what do you see as the biggest risk when investing in Super Apps?

I think the biggest risk is that if you really expect the business model to work, it needs to be a Super App, meaning that it has to be leading across several verticals. Not just one, one is not sufficient to build a big business. The real risk is competition because every vertical would have a number of dedicated players and you really need to provide the service which would be better across a number of verticals. And as well as acquisition costs in terms of acquiring customers, I mean there has to be something better than everybody else in the market in one high engagement service, right? Cracking the one high engagement service first.

Q13. How does your Super App portfolio look like?

Well, yeah, of course. I guess probably the most obvious one is the emerging markets which have a number of verticals open, and a number of digital solutions are yet to take hold. A good example would be in Uzbekistan, it's a relatively, I wouldn't call it a very small consumer market, but it's a sizable market where a number of verticals are yet to be developed.

Q14. Any key trends to watch out for?

The competition is very strong, it's a very competitive environment. I think the interesting thing is that incumbent banks have been lately investing a lot as well to be more competitive with the upcoming startups. I think that companies really need to come up with very differentiated products. On the funding side, 2023 is expected to be the hardest financing markets for tech companies. So, the market will probably bottom down next year. The companies need to be very efficient and kind of be really product focused and make very smart decisions about their spending on their consumer acquisition.

Q15. And that brings me to the final point—do valuations matter?

The valuations they have been, I mean last year they have been very high. This year they are adjusting. I think the private valuations are yet to be adjusted to the current environment. But the valuations are reflections, so the reflections are expectations, how big a business can be, right? Our focus really is not to invest at the cheapest possible amount, but the focus is on building as large a business as possible and kind of freely earning the return from the growth of the business. So, I guess you know it's all in terms of expectations of how big the outcomes could be. And yes, today the valuations are suppressed substantially because of the interest rates. With the high interest rates, any asset which produces potential cash flows, a good number of years from today is substantially negatively impacted from the interest rate rises.

Urs Bolt has over thirty years of experience in wealth management, investment banking and related technology businesses in Switzerland. His core expertise is to develop and roll out new digital business platform.

Q1. Based in Switzerland, how do you view the comparison between the East and the West for Super Apps' growth story?
In Switzerland, we don't have Super Apps. And I'm not sure if they will ever become such Super Apps. If in one segment, like banking, you look at, let's say, UBS online banking, they have only two features, but the user experience compared to most apps you see today, even though Revolut is also not as good any more, as it was, eight years ago or so, because they now have a lot more features. But UBS is like the flagship across and beyond Switzerland, so many functions and features, but you have to search for them. That's banking, right. But then when the call goes across, you have the TWINT payment app, which was basically founded by the banks. And they tried to go in this direction. But I think there will be a trend, but it will by no means beat what you see in emerging markets. I would say it's interesting to follow maybe the Brazil market as well, because they have a similar environment. It's a huge country with 220 million people. So they also have an underbanked and underserved population. But in Europe, I don't see any content when it comes to their demographics, which are against it due to their advanced economies. You also have a much older population. My parents don't go to the internet, I mean, they had it, but they gave up because my brother's living a few hundred meters down the road, so they get it done through him. And it's really best for them, to be honest. It's even a luxury if you don't need it, and you don't rely on the internet. I mean, I feel like it will be huge, it could be a quality of life improvement for me personally. But I don't think I can just give up on that. And so that's why there are many reasons why the Super Apps came or

started to really proliferate and also made a name across the globe, but originating in Asia.

Q2. I should have asked this question before; how do you define a Super App? Because it is different for different people.

First of all, obviously, the hardware is the smartphone. That doesn't need to be the case in the future. And you have one app that integrates a lot of services across one specific use case cluster, like banking or payments. So it's connecting, it starts in one area, let's say mobility, or delivery or ride-sharing. And then it proliferates quickly into payments, which they can integrate if you have a large enough population, then you have your payment services, which dominate like we saw in WeChat Pay and Grab etcetera and, then it extends, they can go into food, they can go into shopping marketplaces and e-commerce. Gojek started more on the e-commerce side, Grab more on the ride-sharing and food delivery side, then you obviously add loyalty points and things like that. And I think Grab is also now having the investment part. So you can start with small amounts to save money. So it's similar to what Revolut does as well. But Revolut is not in mobility, it's not doing ride-sharing, it does a little bit of travelling now, I'm not sure if that's very successful for their large customer base, and will they become successful in Asia? We'll see. But it will be a nice benchmark to look at how they do as a leading FinTech app. So the definition is clear, it's a one in all, or an all in one app, which connects services across the basic needs in your life. That's where it starts. And that's how we can cross sell and scale. So it has the lowest common denominator somehow. When it started, I don't think the inventors of Grab and Gojek, etc, thought about creating Super Apps. It just developed, it's a natural evolution. It's just a very fast evolution, which has happened over 10 years.

Q3. And my next question is on data. Since Super Apps absorb and make use of tons of data. My question is twofold. What do you think of the data analytics, machine learning and the AI? And how do you see that data privacy playing a role? Europe has stringent data privacy laws and in India, people are willing to give away data easily.

Here we talk about data, they don't have the same thinking there. I mean, when you look at how my partner, she was in Dubai for thirteen years, you know, all her and my sensors on our smartphones, we have started to block, all the sensors are off for me as well. Now, usually, I cannot even take a photo, I have to switch it off to take photos. I started this just in December, because I just realized even though I might sound like a paranoid guy in Switzerland, honestly, regardless of whether we have strict data, stringent privacy laws. When you look at the legal environment, that just gives you a sense. The data is being used anyway, by secret services, FBI and Homeland Security in the United States. Because they have all these legal frameworks in place. You call it Anti-Terrorist laws. These people even voted yes, because they didn't really look behind the lines. Last year, my son voted clearly no, he understood that it goes too far. Because anyone, including me, if I'm let's say a citizen rights advocate, which I am in the hidden, you don't see that on my X profile. I have a second profile for this, because I had an incident last year in May and it was shocking to me because you cannot open your mouth. And I'm not the person who attacks people and confronts them like some others on X. So I had to change that. But the fact is, anyone suspicious could be called, 'Okay, we have to investigate that'. Now, if these anti-terrorism laws that are across the world now, and the same is happening in the Philippines, you can actually use personal data and observe people. Now in these emerging markets, the people are so used to social media and sharing their secrets about their private lives, anything and everything is shared. I now have a god child and I realized that he has ten godparents, not just one godfather, which was interesting to me. And it was funny, they shared everything. The baptism, of course, that's a nice and positive thing but, I mean, they do not have the same feelings about data privacy, like we have here and in Germany, and some other countries. But most people in the West also don't realize it, I would still probably call myself a bit naive. I lost a lot of trust in the government, when you go to Southeast Europe, where there was communism where the bank accounts were suddenly all blocked. The government confiscated it, dating back over two decades, with my ex-wife. And they don't trust the government as much there now as we do here in Switzerland. And I'm wondering how it will be when they realize what's going on behind the lines. Now in these developing markets, where the informal economy is so big, and you have to make a living out of it, I think it's also a necessity. They also benefit from sharing the data with each other. But if you automate the system, like these big tech companies like Grab and GoTo, and all these Chinese companies. They are big tech companies, It's part of their game, you know, to push out these services, and to keep the flow of interaction, whether it is informal exchange of information, or advertising, or pushing for commerce. It's just part of it. When you look at emerging technologies, it all comes together, automation, robotics for delivery or preparing deliveries, etc. The Internet of Things (IoT) with all the sensors, and AI, based on cloud computing because everything is running on the cloud. It is high performance, of course, you depend on these cloud technologies, which is one of the big, new risks we have with all these big tech companies running on the cloud. And then you have decentralized approaches with distributed ledger tech blockchain and smart contracts which are going to automate this and make it super-efficient. That's how you can actually include the lower part of the society and the economy into this Information Age, which in the

West, we don't know how to deal with. Switzerland has no idea how to do this, even though we have super geniuses, like myself, you know, at ETH, and at Google which has, like 8,000 people in Switzerland alone, the biggest office outside their head office in California. But we are not the retail market. And even if you have Germany which has 80 million people, the UK 60 million, France around the same. They don't have the scale of what you have in Indonesia, India, China, but also the Philippines where you have a lot of underserved potential which can go into the market. But you need all these technologies, because you cannot do it manually. You know, people always thought before, you have cheap labour, you can do it manually. That's why you're so cheap. No, the thing is, in this digital age, you need to automate this. Because you cannot do that with labour. It's error prone, it's too slow. If you go to WeBank, you will get a loan for an SME within seconds. I still haven't come across any such digital bank in the Western world. I just haven't seen it.

Q4. Grab and GoTo enjoy a unicorn status and have high valuation but not making profits yet. How do you reconcile the two- valuation skyrocketing and still not making profits?
Obviously because the perspectives are that the sky is high. It's like a fantasy, right? It plays into these huge growth markets. Now they get large, huge corrections. And you see that already with consolidation, which has started. So GoTo came together, it was Gojek, Tokopedia. There was something else which was integrated as well. Another company, a third one. But let's leave that aside. There will be consolidation. Of course, it's like Amazon in the West, right? They are the market leaders. They were the first, and how long were they in loss? I think it was only 2013 when they turned into the positive territory. And people still bought their shares. So it can take a very long time. But, I think consolidation

will take part and market share is important. And, from there, they will start going to higher value services and products and start earning fees from it. Because the growing middle class, which is obviously much lower in terms of numbers, per capita per person than here what you would deem middle class, especially in Switzerland. But I mean, that's also one reason why it's good to go to the Philippines for me, it makes life much easier. But they will have to start earning money on probably the value added services. And for those who want to earn money on the basics, with margin business, etc., they have to become even bigger.

Q5. And do you have perspective also on the business model of these apps because you are teaching this in universities in Switzerland and around the world?

The business models are developing, I mean, with the growth, I think the ones which cannot adapt their business models to changing circumstances, they will probably either exit or reduce to the profitable segments instead. And then they might be taken over or they might be successful in that niche. But otherwise, it goes with technology. It's all very tech driven. Most of these founders are engineers, or a team, with combined intelligence, it's very tech driven so I think it will go along with tech. And because there are less hurdles or lower hurdles to using these technologies in these markets, It will also proliferate faster than, let's say, maybe in Europe. I mean, in the States it is also easier in a way. But some states in the United States have strong laws. California is more like the Germany of the States, similar in their laws and how they deal with some policies in anti-, etc. Which is not really positive in my view. And they copy that. But that's their problem, not mine. Now, the business models just have to evolve, I don't know what they will end up with. I mean, for me, it's just fascinating to follow these very dynamic spaces. And look at it from that point of view, compare to what's going on in the rest of the world.

Q6. Let's say if we catch up after five years, do you think we would still be talking about Super Apps?
They might have another buzz word for it. Could be you know, it could also be a mega app or so. Yeah, I mean, I think I took up that term very quickly and easily. You can explain it in various depth and people understand it. Because you can also take your smartphone, it can open WeChat, even me as a foreigner, I can look at so many services on it, or you can go to the Grab app. I still don't delete this app just to look at it. It's just interesting, even though you might not have access to them, because especially in the financial spaces, it's all you know, local regulation or country specific regulation and you need to somehow prove your identity, you cannot just open it because you're bound to the banking law, usually. But you can still see the app icon. And the rest, you can look up and read about online.

Q7. Do you see Europe playing catch up?
I think it will not be the same way we look at Super Apps what we discussed, for Europe. What we need is process integration, so offline to online, and integrate that into the complete customer experience. I see that in the B2B side in the competitive areas, I think about Bosch from Stuttgart in this case.

Bosch is something which still comes across, still very noisy, very process driven. And a lot of Swiss firms too, let's say, I would assume Geberit toilets systems. It's a Swiss company near where I grew up. Everyone has these toilets everywhere in the hotels and homes, probably going to have one of those in my home in the Philippines. You can get them all over the world, and they have a global business, they need to become more efficient. So they will, along the processes, they will integrate these processes into apps. But there are more segments, specific verticals, specific Super Apps, then, you know, to integrate suppliers, service contractors, and even processes. And I could imagine that will happen also in the financial space. But there it's about being value added. And they always want to see a business use case. And I don't think you will call these Super Apps. It could be like a private marketplace for private investors, let's use UPS, you know, they have a lot of features and functions with a lot of legacy technology, they will never be able to clean up this legacy technology, the way they operate and manage their business. So they will just integrate apps, somehow connecting them, and hopefully get a more or less similar user experience or a better one even. They don't have the same, let's say conditions like you have with Super Apps, which started in the retail space micro businesses. And only green field with small use cases, but to see how complex Super Apps look like and the user experience of WeChat, to me, honestly, it's not superior. People just got used to it. But the user experience of how you handle and deal with it was in a way, shockingly low. I wasn't really convinced of that, I was more astonished by the variety and all the services integrated across your whole life, and needs you have as an individual.

Q8. Indeed, I think you're right, because it's just going to shape up in a different way you may see Super Apps, but the form and the shape would be very different from Asia, because the demography, the kind of the economic growth that we see in these two contrasting markets, I think it's just not going to be the same way.

I think, like in ten years, maybe part of the services will be maybe edge computing, outside the apps, but connected. But the smartphone will be maybe like the flagship carrier or the aircraft carrier, like sort of, you know, and you have like IOT, I mean, you already can do it today, right to connect your security system of your home, together with the cameras, or I think it will just there will be a bigger variety of technologies being integrated into the Super Apps. And that's a natural evolution, it will not be revolutionary, people just get used to it. So it will be VR. It could be I don't know what the metaverse will do, you know, I mean, I'm following this too. For me, it's just a virtual ecosystem. But it's very early. But augmented reality could be interesting for many use cases. Then you might have very specific hardware pieces, which are connected maybe to the carrier to the Super App, as such, you know, via 5G Internet of Things. Then, of course these things can only work with AI, cloud computing. You might even have a chip in your hand. You know, I mean, some people do that now. You read about it like for payments, or you have a hardware wallet, you know to be your own bank. What else I mean, the fantasy is boundless. You have the technologies around, in one way or the other. And if you look at Frontier Markets even more than emerging markets, a metaverse really cannot work.

I mean, you're gonna be dependent on Elon Musk's Starlink. Even in the Philippines, if I want to do lectures, and I need top quality, I might need to get the Starlink service, which are satellites that belong to a big tech entrepreneur. But this is all coming together. Also, you will need edge computing and IoT, just to reduce the dependency on the central cloud services. And then think about quantum computing when that comes in. But then you need the connectivity to where the services apply, think about offline to online connection. So there's still a lot to do. In terms of what's all around, you could already build it up if you just build. I mean, if you have enough money and you're willing to pay for it, you can build it all today.

Yeah, you're right. It's always a two-way traffic

So this is something which in my view is very important to communicate, that you and me have the opportunity to put it into the media and communicate to just make sure that people aren't just blind and naive and following this. That's why I'm also holding back with a lot of initiative push with buzzwords. Just the keyword ESG (Environmental, Social, Governance) and greenwash. There's a lot of greenwash, there's a lot of narrative pushing. And when you scratch the surface, you see it's a lot of communication, but not much behind it sometimes. And also the big businesses at the end, push away all the well-intended initiatives from startups possibly. We need to make sure it will be equitable, and it will move in a way that society benefits and not just the few who have the power and the means to take over the market. And that's very difficult when you look at history, to assure the city cannot rely on the court of law. And then what's democracy? What are free markets? I don't think we have free markets any more in the West. It's maybe not the command and control economy like you had in communism. I mean, even China is not this far away from what you saw what happened in the Soviet Union, etc. But still, when you look at how many buildings they build, all these cities with empty sky high buildings. It's also quite amusing, right?

Keegan Sard is a fintech consultant based in Monaco.

Q1. Being based in Europe, do you miss using Super Apps?
We don't see any apps in Europe, like, literally, it's nothing over here. Europe is like 1995, when it comes to technology, it's god-awful. You know, you're lucky in Europe to get access to an app that allows multi-language. And that is sad, that is terrible. My bank said to me, unless I'm a private customer, and giving them a minimum balance of something absurd, they won't turn it on for me. It's the same app. It's the same infrastructure. It's literally a tick box, and they won't do it. So I find it really funny, we're talking about Super Apps here, but in Europe, there are no apps at all. The app landscape is god-awful. It's not like Australia, the US, the UK, Asia, you know, it's not convenient, it's not intuitive. There's no UX/UI, they just don't care.

Q2. How do you think we can change that? Do you think something needs to be fixed there? What can be done?
Well, you know, my view on this is for the government to get out of the way. It continues to put barriers to entry that are just blockers to innovation. You know, it's all good that we go to Davos next week. And we talk about innovation and technology and climate change and all these big issues. But then they don't do it. Because they want their cake and want to eat it too. It's not looking at it from a position of saying, right, we can solve that problem. A great case in point is, you know, you could look at a cashless society using apps, but we don't do it. You know, we could look at ways to make people's transportation better. Monaco just tried it. Free public transport lasted three months, and they're back to tickets again. As a consultant, I would love to see the delta around the costs involving having ticket machines and compliance officers and all this stuff that limits transportation options, mainly for people that are less well off as well. And we

continue to put these blockers in. So you know, I'm a big proponent of Super Apps, if they're given the allowable rope to succeed.

Q3. COVID-19 has been a big accelerator for the cashless society and bringing the offline-to-online switch. Do you see any kind of surge in Super Apps, especially when you come to Singapore or Indonesia?

Well, the best is WeChat and Alipay in the Asian region, for sure, by far. The closest thing in Singapore would probably be Grab, you know, takes in transportation, and takes in food and takes in a bit of everything. But my fear is that with Central Bank Digital Currencies (CBDC) coming in, in the next sort of twelve–eighteen months, it's going to hurt the Super Apps market even worse. And you know, the move to cashless is supposed to be more liberating and more free, but it will be less free and less liberating. Because you just need one government to turn around and say, 'Oh, yeah, we will never stop you from buying McDonald's five times a week' and then all of a sudden, one government tries to get away with it. And we saw this during the pandemic, which is a great example of this.

Q4. You spoke about CBDC and it is a very interesting and hot topic, especially China has taken leaps and bounds. So, when we talk about CBDC we also look at the financial inclusion aspect. Do you think Super Apps could bring that kind of a balance by serving the unbanked? Especially because this is the mobile-first experience for them?

Look, the change management process of this is going to be very interesting. You know, especially amongst the older generation, and the poor, and I mean, not just the poor that I mean, the poor that are homeless, right? At least cash gives them an avenue to participate in society. When you go cashless, well as it is, you know, there's a number of jurisdictions, you can't get a bank account unless you have an address. It's absolute insanity. You know, you can't get a bank account unless you have an ID, but you can't get an ID without money. So, you know, and then you can't sometimes get an ID without an address. So I found it quite interesting. When I moved to Europe. It's like, oh, you need this, but I can't get that until they have this. And it's just this cycle of like, 'oh, okay, I got it'. 'Yeah, cool, go here'. 'Oh, no, you need this document', and I can't get that document until I get the bank account, I can't get the bank account till I have the business, can't get the business account if I don't have a bank account. It's insane. You know, organizations like the World Bank, and the UN should be focusing their efforts, on these short, small wins, in my opinion, you know, like, encouraging governments to make it easier, you know, with Anti-money laundering (AML) and Know Your Client (KYC) issues that continue to plague our societies, we're not stopping the problem. If you're in the banking system, you're fine. You know, like you don't get flagged for the transaction, you should get flagged for, you get flagged for the bullshit lunch that you had, which was a legitimate transaction. And it just blows my mind. You know, I can be paid by a new company with no issue at all. And then it just goes sailing through and then I have lunch somewhere, and they're like, 'Did you have lunch today?' It's insane. So I do feel for those lower, less fortunate people in our societies that once we go cashless, get left behind, again, when they shouldn't be. You know, one thing Australia did well is I can get a bank account with a mobile phone number in about three seconds with pretty much the majority of banks now in Australia. I can't do that in Europe. And that's all because of the EU legislation around AML. Because they think that we are all dirty criminals, instead of looking at the 1 per cent that actually are. Super Apps are going to have the same problem because they're going to have that financial mechanism inbuilt into them.

Q5. How would describe the Australian Super App market?
No, there are seven different apps in the same industry. There's no gold standard for Super Apps. The one thing Australia has done very well, and it should be adopted globally is the NPP, the new payments platform.

So the RBA, the Reserve Bank of Australia put out a set of guiding principles and technology improvements to be done by February 2018 or 2019. But they have allowed for near real-time payments. So I could pay you with a PayID, and PayID does not have to have a numeric value, it could be a mobile number, just has to be unique, it could be an email address. And you would receive those funds in real-time or near real-time, microseconds. And that has changed the banking system for the better because you used to have to wait days. I don't understand why we have bank days any more. Like technology doesn't stop. Why do we allow it in Europe for five, or six days to get access to your money? We are not in 1962 any more.

Q6. So when you wear the Venture Capitalist (VC) hat I'm sure you're looking at Super Apps in a very different way. What are the things you look for?
What I look for is, you know, is there a generic need? Like, is there a specific need for the Super App? Can it solve that problem? Right now, I haven't seen it in the Western world, like I think I use the example of X being a potential one, because at least they have a large user base. Meta before their challenges in the stock market, in recent times, was also looking at this as a payments platform because they have grown in users, I don't think you can do it without growth of users, I think it's too hard. You would need, you know, for example, I don't understand why Uber hasn't merged their Uber Eats and their Uber driving apps as one, that's a logical thing for me, instead of having multiple apps, they should just be one like Grab had done quite successfully.

In regards to payment options, like you've got cloners, and AfterPays, and all these different payment providers now. But if you look under the surface, you still have Visa, MasterCard, and American Express rails. So at the end of the day, you are looking at the same triopoly of the same providers offering the same COBOL (Common Business-Oriented Language) back solutions that are just rubbish. In regards to valuations, very hard to put a number on it, I think you would have to come to the market, like I said, with a strong user base that you could convert, you know, Apple is doing it with Apple cash, you know, they've got their Apple card now, you know, they are in the Box Seat, you know, Apple Pay is now accepted in several retail fronts, not just online, but on mobile. That makes sense to me. Could they go further? Sure. You know, there was a rumor at one point that they might look to buy Tesla. Now that would be interesting, because Tesla's sort of grand vision is that it will do autonomous vehicles for pickup and drop off like Uber. If they went down that path, then that's now starting to be a Super App, right?

You know, I always joke that when the bank calls you and they ask you to confirm your identity. And I'm like, this should not be allowed. It should be in an app, it should be like, 'Hey, I've sent you a push notification in your app. What's the number?' and I go, 'Oh, it's 1-2-3-4'. 'Okay, you're Keegan, done'. So they're the challenges we're going to have with Super Apps as well.

Q7. From adversity comes creativity. Do you think apps like WeChat and Alipay in China were born as a result of it?
China wants to track everything. So, like, it makes sense for them to support a Super App ecosystem. I wouldn't be surprised if WeChat and Alipay have back ends into the government already. I don't know if that's confirmed, but I wouldn't be surprised.

Q8. Let us say Super Apps take over the West. How do you think people would respond to data collection requests? And what do you think people's views on the data collection practices would be because they collect everything from like the time you come home from work and, and things like that your credit line and things like that?

I can give you a great example of TikTok. In April, 2023 Tik Tok was banned by the US military, the US government, and Congress. The state of Montana became the first state to sign legislation into law to ban TikTok, because they were tracking everything, keystrokes, they were tracking pretty much your entire life. And as more people understand Terms of Service, you know, I would love to see Terms of Service in one page, I think Apple did a great thing with iOS 16 around just what is being shared under privacy on the App Store. So you can see that in a really easy, understandable language, there was a guy called Alan Siegel, who did a really great TED Talk back in the day called 'Understanding legal jargon'. His challenge was to grab the world of data around credit card terms of conditions and put it onto a single page. And all these lawyers said it can't be done. He proved them wrong. So I think we need to come down to that level as well, because as soon as the West understands, I think there's going to be a shift in privacy. I think COVID has really proven to me that it was mishandled and more and more is coming out about that over the weeks, months and years, there will be a greater shift to privacy. And you'll see governments having to react to ensure that they get reelected.

Q9. Yes, privacy is definitely a key issue across the world. So in terms of Super Apps, what are your views on being centralized versus decentralized?

I think everything has to be decentralized. You know, my utopia is where the government is seen and not heard. We need to really focus on the 1 per cent of problems and not treat everyone like the problem. You know, that would be the idea. I think we've got some great technology on the blockchain. It needs to be utilized more. I think, if you know, Ray Dalio and others are right, we're going to have a shift in the reserve currency this year, to potentially BRICS, which is Brazil, Russia, India, China, and South Africa as a global currency backed by Gold. If they do that, the economic downturn and consequences will be magnified 10x, 100x. And the only thing that I think we should be going through is decentralized for sure.

Q10. Yeah, definitely. So speaking of the decentralized future, how do you think the future looks like? You know, he spoke about X becoming like, it could become a Super App or Meta. So what do you see coming?
Look, I think right now there's three sides, there's the cash economy, which, you know, arguably saved us from the biggest recession we would have had in 2008. Then you've got the card economy or the online economy and then you've got, you know, the cryptocurrency-type space. If you remove cash, I still think two remain. I think blockchain is going to be here to stay. I think cryptocurrency has decided there will be a segment of the market that just says we're not participating in your CBDC environment. And then it's like, how does that work? How do people trade crypto if you know the banks turn you off? Because you know, it's one thing to get your money into crypto, but it's also very hard to get it out.

Q11. Do you think we will be still talking about Super Apps five years, ten years down the line?
I think there will be a Western Super App at some point. But I think it'll be an Apple or Facebook that creates it based on their payment infrastructure, because I think payments run the

world and if you get payments, right, everything else will flow from it.

Q12. Any future predictions?

I think 2024 is going to be interesting because every company is forcing people back to the office, especially the large ones, you know, they want you in the office three, four days a week, now, you've seen a shift in the, you know, we now see the four-day workweek, that that is an interesting shift. However, when we look at it broadly, we have a situation where it's going to go again, in two streams, it's going to go from a remote-first culture, and everyone's going to jump on the bandwagon and go down that path. But then you've got countries like Switzerland that don't allow you to work outside the country. So, you can't work for a Swiss firm and be in Monaco, for example. So, you know, you need that regulatory side to actually understand that the world is a very different place to when these laws were originally enacted. You know, I love the rule that if you bring in a new law, you should take one away. Because we all have stuff that just isn't relevant any more, or isn't suitable to our current society, our cultural norms, etc.

Q13. What's your view on the data monetization?

I love AI. ChatGPT is going to change the world.

It's an addiction that I am there every single day. If they said to me, I need 1,000 bucks a month, I'll find it. Like, I'd find it for sure. It's like, I'll sell everything, literally live in a suitcase, or a sleeping bag and just have my ChatGPT run the world, right? Like, it's, I'm going to put out a bold statement here. And you know, call me in ten years if I'm still alive. And it will be that AI will change jobs forever. AI will take away jobs. But it's for the better. And I truly believe that data will not be sold in its current form and people will be paid for their data, not allowed to use a free software tool in order to give away personal data. I see X one day paying me. So that's how I'd like to think we're in the age of content now. And if you're producing content giving and content is data, right? Like I'm just telling you where I am or what I'm doing. I think you can in a new world be remunerated for that, for advertisers.

Q14. We've talked about ChatGPT and emerging technologies. I'm curious to know whether you think these developments are a cause of worry? Do you think that maybe they're enabling us in giving too much away to the point where it feels like you're kind of selling yourself?

I think we have one life. And if AI allows us to have more time spent with the things that actually truly matter. Your career doesn't matter. It's you know, like, if I said to you today, would you do your job for free? 98 per cent of people 99 per cent of people would say hell no. You're not there. You know, they're like, I love it when people like, oh, you know, when they retire, it's like, oh, you know, my profession, and this and it's all bullshit. At the end of the day, it's all bullshit, because you wouldn't do it for free. At the end of the day, if we can solve problems with technology, and move to a, you know, universal basic income, for example, I think we could all live better, happier, healthier lives. And, you know, go back to the things that matter.

Q15. Right, so my last question to you is, if you want to bet on one Super App, which would you bet on?

I think X or Apple will be the first to market. They're the two that I'm most excited about. I trust Apple more than I trust the government.

Q16. Any top picks from Asia?

Well, the one I used in Asia was Grab, and I think Grab will continue to expand and grow. That's probably a model where

I think Uber made a mistake. And on that with a couple of seconds to go, Uber made the mistake of selling their infrastructure, but forgetting that people travel. So Uber could have been that Super App. So yes, it's owned by Grab now owned by Didi or owned by someone else. But as a foreigner in a different country, I should still be able to use the Uber app, because I'm an international person. You know, when I first went to China, Uber saved my butt, because like, it's English. So I was able to travel for the first time ever without a translator. Now, on the flip side, now, when you go to China, well, they fixed this problem. Now Didi made you have a Chinese credit card, and it was only in Chinese, and a Chinese number. So as foreign tourists, I couldn't use the app. And I think that was again, an inherent inhibitive of a socially inclusive system, as I mentioned with banking and finance earlier.

Tram Anh Nguyen is the co-founder of CFTE, a global impact platform for education in Fintech and the future of financial services.

Q1. Super Apps are gaining popularity across the globe, especially in Asia and emerging markets. What are your views on them?

The growth of Super Apps in Asia and emerging markets is rapid and viral. I think there are several reasons behind it:

With recent advancements in infrastructure, mobile networks, logistics, online payments, we have seen significant inflection in digitalization in emerging markets. In addition, the increasing mobile phone penetration in developing countries enables more people to access the digital services.

Nowadays, Super Applications like Grab, WeChat, and Go-Jek virtually completely cover every element of a user's life in Asia, whether it's booking a taxi, making a restaurant reservation, or paying a bill, etc. For example, Grab was chosen by more than 90 per cent of the respondents in Malaysia.

I hold a positive view of this situation. With the rise of technology in finance and e-commerce, there are more Super Apps developed and adopted in emerging countries. It can create a more convenient, smart, and inclusive ecosystem for the community. However, we still need to take data regulations into consideration, especially in developing regions where the regulations are not strict and structured. And in developing countries, there is a digital gap between users and developers, so there are still many things we can do to bridge the gap.

Q2. Why are Super Apps more popular in the East than in the West? Are there any cultural, regulatory, or market-related factors that contribute to this difference?

Regional differences in resources, public policy, and customer tastes produce different approaches to new technologies and innovations, which can result in various outcomes. I think the main differences between the East and the West of the super apps are the regulatory landscape and economic component.

- **Regulatory conditions**

  Although apps that fit the super-app classification can offer users a wider variety of services in comparison to single-purpose alternatives, internet regulators in regions such as the US and Europe have become more concerned about the overall power of the technology industry. European countries have become more critical of companies developing Super Apps as one characteristic of Super Apps is that they depend on user data and third-party connections.

  The majority of Asian nations lack uniform laws and regulations governing how big digital companies utilize the data of their users. On the other hand, Europe takes data privacy seriously. For example, the framework for the General Data Protection Regulation (GDPR) demonstrates how seriously Europe takes this issue.

- **Eastern economy is dominated by SMEs**

  SMEs continue to be the foundation of most economies in the East, especially in those where Super Apps have a lot of potential to grow as a result of factors like financial inclusion and younger demographics. To fully digitize their firms, SMEs can access all of these features and capabilities via the Super Applications such as Alipay. The more SMEs a Super App hosts on its platform, the less likely its users are to abandon the app because everything they require is contained within it.

  In addition, I think East and West also hold differently on the concept of 'Super App.' Take Apple's iPhone as an example, it's all about the 'services' but not limited to a 'Super App'. You can access everything you need via Iphone, such as Apple music, Apple pay, Apple TV, and even iMessage. The key for Apple is creating a service ecosystem, not an app.

Q3. How do Super Apps leverage data and analytics to drive user engagement and retention?

User engagement analytics is essential for developers as it tells whether users are getting value from the product or not. From my perspective, localization is really important for developing such apps. Super App developers need to vocalize their apps because understanding the local market is essential.

For example, when Gojek first started out in Indonesia, people in Indonesia couldn't afford smartphones, not to mention downloading the Gojek apps. In light of this, Gojek gave incentives to their driver partners to sign contracts with them and provided those drivers with loans to purchase smartphones. Therefore, Gojek expands its market share in Indonesia, or more broadly in South-east Asia (SEA).

Q4. Are there any ethical concerns or challenges that arise with the development and use of Super Apps, for example, issues regarding users personal data collection and privacy?

I would like to emphasize that using the 'right' data is also important, developers need to ensure the personal data they collect follow the regulation framework and need to provide users clear and concise information about how their personal data will be used.

Q5. What role do you see emerging technologies, such as artificial intelligence and blockchain, playing in the development and growth of Super Apps?

We've seen Super Apps evolve developing their payment system within the application. As a result, it is important for developers to have a deep understanding of the current financial industry and banking systems in order to build effective payment systems within these Super Apps. This requires cooperation between tech companies and financial institutions to create a seamless ecosystem for users. In addition, AI can be used to enhance the user experience within Super Apps by providing personalized recommendations, real-time language translation, and predictive analytics. AI-powered chatbots and virtual assistants can also be integrated to handle customer service inquiries and improve the overall efficiency of the app. And blockchain technology can be used to improve the security and transparency of transactions within Super Apps. By using blockchain, data can be stored in a decentralized and immutable ledger, providing increased security against hacking and fraud. Additionally, blockchain can be used to facilitate peer-to-peer payments within the app, reducing the need for intermediaries and lowering transaction costs. By leveraging emerging technologies such as AI and blockchain, Super Apps can provide a more secure, efficient, and personalized user experience, which can help drive their growth and adoption.

Q6. Do you see Super Apps lasting the test of time? Will they still be a large part of our lives in the coming five–ten years? And are they here to stay?
Firstly, Super Apps are often designed with a strong emphasis on user experience, which can help drive user engagement and loyalty. I think regulatory and security concerns are main issues that could limit the growth of Super Apps. Another factor dominates the usage and development of Super Apps is education. Education not only for the community but also for the developers and businesses is fundamental for all!

By equipping developers with the necessary skills and understanding of the latest technologies, they will be better equipped to build and integrate sophisticated payment systems into Super Apps, providing a seamless and efficient experience for users. We also need to educate SMEs to use the new technology, making sure they can grow their business.

In addition, providing education and training on the latest industry knowledge and technologies to the users (community) is essential as well. Collaboration and coordination between government and industry is crucial in bridging digital gaps and improving technology literacy among the general population. Government can play a key role in creating a supportive environment for the development and adoption of new technologies, while industry can lead the way in providing education and training programmes to equip individuals with the skills and knowledge needed to take full advantage of these technologies.

Once everyone is equipped with the knowledge and skills to adopt, develop and understand the logic of the whole ecosystem, Super Apps can still be a large part of our economy.

Q7. What do you think Super Apps will look like in the future, and how will they change?

From a 'connectivity' perspective, it's important to note that **an app is not just a tool, it needs to connect people and create an ecosystem** for the community, businesses, and industries to interact and benefit from. This means that the app should facilitate communication and collaboration between different stakeholders and create opportunities for them to connect and engage with each other.

Another important aspect of Super Apps is localization. Every market is different, and cultural and regulatory differences need to be taken into account to ensure that the app is tailored to the specific needs and preferences of the local market. For example, a Singapore model of a Super App may not be successful in Vietnam as the market, culture, preferences and regulations are different. This involves conducting market research, working closely with local partners, and adapting the app to the specific needs of the local market.

Last but not least, in the future market, customers will increasingly value **corporate social responsibility** as consumers grow concerned about the quality of life and environment. How developers embed sustainability and inclusivity into their business model will differentiate their products/services in the industry.

Final Thoughts

Hope this book on Super Apps has provided a comprehensive examination of the evolution and impact of Super Apps on the world of money and technology. The first part of the book traced the history of money from paper currency to digital money and how it has shaped the way we think about money today. The book delved into the cultural differences in the adoption of Super Apps, with a focus on the East and West. It examined the reasons for the popularity of Super Apps in the East, such as the high penetration of mobile phones and the lack of developed traditional banking systems. In contrast, the book explored the hesitancy in the West, where concerns about privacy and data security are more prominent. The book also looked into the potential for Super Apps to break into new industries and disrupt traditional business models. For instance, Grab, a ride-hailing app that has expanded into food delivery, grocery delivery, robo-advisory, and financial services. The book also explored the potential for Super Apps to aid in financial inclusion, particularly in developing countries, by providing access to basic financial services to those who may not have had access before. Examples of this include PayTm in India and GCash in the Philippines. The book also looked at the impact Super Apps can have on society and how they may change the way we interact with technology and each other.

The second part of the book included in-depth interviews with experts in the financial industry, providing valuable insights into the current state of the industry, as well as predictions for the future of Super Apps and their impact on the financial world. For instance, Bakhrom Ibragimov, a Venture Capitalist, shared his thoughts on engagement and cross-selling, the honey pot of Super Apps. Governor Serey Chea of the National Bank of Cambodia shared insights on regulatory practices in the Cambodian market. Venture Capitalist Keegan Sard provided his vision for the future of the Super App industry. Fintech Influencer Chris Skinner offered unique perspectives on the cultural aspects of regulation. Professor Patrick provided a consumer's viewpoint etc. Through these conversations, the book illustrated the perspectives of key players in the financial industry on the rise of Super Apps and their potential to disrupt traditional business models.

Overall, this well-researched examination of Super Apps provides an illuminating picture of how they have evolved, what they mean to different cultures, and how they are shaping the world we live in. It provides us a unique perspective on the industry, giving us access to the thoughts and insights of some of the most prominent figures in the financial world. It provides a more complete understanding of the current state of the industry and the potential future of Super Apps. Looking ahead, it is clear that Super Apps will continue to have a significant impact on society, changing the way we interact with technology and each other. As technology advances and more people gain access to mobile phones and digital services, Super Apps will play an increasingly important role in shaping the future of finance.

The impact of Super Apps on the world of finance and technology is only set to grow. The insights gained from our research and interviews have shed light on the potential for Super Apps to transform industries, break into new markets, and provide valuable services to millions of people around the world. Super Apps are poised to play a significant role in financial inclusion efforts, particularly in the developing countries, where traditional banking systems are often inaccessible or underdeveloped. With their ability to provide access to basic financial services to those who may not have had it before, Super Apps have the potential to empower millions of people to take control of their financial lives. As we look towards the future, we see a bright path ahead for Super Apps as they continue to compete with established players like Grab and GoTo and expand into new markets. The potential for Super Apps to disrupt traditional business models is significant, and their impact on society and the way we interact with technology and each other cannot be overstated. So let us keep watching as Super Apps continue to shape the world we live in. The potential for innovation and growth in this industry is immense, and we are excited to see what the future holds.

As we close the book on Super Apps, we can be sure that they will remain a fascinating and important topic for years to come. Soon, Super Apps and Fintech will become integrated into every part of our lives. As Chris Skinner said, 'Any sufficiently advanced technology is indistinguishable from magic', and we look forward to seeing that happen with Super Apps in the Fintech space.

. . .

References

  i. https://en.wikipedia.org/wiki/The_Travels_of_Marco_
Polo

  ii. https://www.reuters.com/article/grab-results-
idINKBN2770YV-

 iii. https://www.reuters.com/article/cbusiness-us-grab-
southeastasia-exclusiv-idCAKCN1VG003-OCABS

 iv. https://www.itu.int/hub/2021/11/facts-and-figures-2021-
2-9-billion-people-still-offline/

  v. https://www.statista.com/statistics/262966/number-
of-internet-users-in-selected-countries/#:~:text=As%20
of%20January%202023%2C%20China,around%20311%20
million%20internet%20users

 vi. https://www.businessinsider.com/what-getting-fired-
looks-like-in-different-countries-2019-7

 vii. https://phdessay.com/hierarchy-and-power-within-east-
and-western-enterprises/-

viii. https://www.omfif.org/2020/11/chinas-winning-cbdc-
approach

 ix. https://www.npr.org/2019/10/30/774749376/facebook-
pays-643-000-fine-for-role-in-cambridge-analytica-scandal

  x. https://content.11fs.com/article/serving-the-unbanked-in-
southeast-asia

 xi. https://journals.plos.org/plosone/article?id=10.1371/
journal.pone.0253568

Acknowledgements

I am grateful to my team at FemTech Partners[1] for their unwavering support and for making this book possible.

My gratitude to the industry veterans for making themselves available for the interview; your insights added a lot of value to the book. Thanks to Serey Chea, Governor of the National Bank of Cambodia; Sameer Nigam, founder of PhonePe; Janet Young, MD & Head, Group Channels & Digitalization, UOB Ltd; Chris Skinner, Fintech influencer and author; Phanindra Sama, Angel Investor and co-founder of RedBus; Pun Chi Seng, Associate Professor, Director of MSc in FinTech, Nanyang Technological University, Singapore; Tram Anh Nguyen, Co-founder of CFTE; Bakhrom Ibragimov, Venture Capitalist; Keegan Sard, Fintech consultant, and Urs Bolt, Finance professional.

My utmost gratitude to Amberdawn, my editor, Nora, Swadha, and the publishing team for their steadfast support and guidance.

Lastly, to that friend who proposed the book idea but walked out. I am so glad my persistence paid off; I am a published author and realized I am enough.

[1] Femtech Partners was founded in 2019 by Neha Mehta when she realized the disparity in the number of women present in the FinTech industry. She decided to quit the corporate world to create a level-playing field for women and flatten the curve. The firm specializes in FinTech, financial inclusion and promotes women in the finance industry.